S0-BSN-252

Interpreting Basic Statistics

A Guide and Workbook Based on Excerpts from Journal Articles

Third Edition

Zealure C. Holcomb

Pyrczak Publishing

P.O. Box 39731 • Los Angeles, CA 90039

Although the author and publisher have made every effort to ensure the accuracy and completeness of information contained in this book, we assume no responsibility for errors, inaccuracies, omissions, or any inconsistency herein. Any slights of people, places, or organizations are unintentional.

Project director: Monica Lopez.

Cover design by Robert Kibler and Larry Nichols.

Editorial assistance provided by Sharon Young, Brenda Koplin, Cheryl Alcorn, Randall R. Bruce, and Elaine Parks.

Printed in the United States by McNaughton and Gunn, Inc.

Copyright © 2002 by Pyrczak Publishing. All rights reserved. No portion of this book may be reproduced or transmitted in any form or by any means without the prior written permission of the publisher.

ISBN 1-884585-39-6

Table of Contents

Continued →

Notes:

Introduction to the Third Edition

This book presents brief excerpts from research journals representing a variety of fields, with an emphasis on the social and behavioral sciences. The questions that follow each excerpt allow students to practice interpreting published research results.

The questions require students to apply a variety of skills including

1. locating specific information in statistical tables, figures, and discussions of results;

2. performing simple calculations to determine answers to questions not directly answered in the excerpts;

3. discussing the authors' decisions regarding reporting techniques;

4. describing and interpreting major trends revealed by data, including evaluating the authors' interpretations; and

5. evaluating procedures used to collect the data underlying the statistics presented.

Although the excerpts emphasize the "results" sections of the journal articles from which they were drawn, some information about procedures, such as sampling and measurement, is also often included in order to put the results in context.

Some Assumptions Underlying the Development of This Book

A major assumption is that students will find materials based on actual research reports inherently more interesting than the hypothetical examples typically used in statistics textbooks.

It is also assumed that students will benefit by practicing with materials written by numerous authors; this allows them to see variations in the uses of statistics and in reporting techniques as they are actually used by practicing researchers.

Finally, it is assumed that a collection of complete research reports would produce too much material to be integrated into traditional statistics courses. Instructors of such courses are often pressed for time when covering just the essentials. Hence, this book presents brief excerpts to conserve instructional time.

Cautions When Using This Book

Students should be aware that the exercises are based on excerpts from journal articles. Although the excerpts are in the original authors' own words, many important points made in the complete articles were omitted for the sake of brevity. Before generalizing from the excerpts, such as in papers written for other classes, students should read the full articles, which are available in most large academic libraries.

Although answers to the *Factual Questions* are either right or wrong, there may be more than one defensible answer to each of the *Questions for Discussion*. At first, some students are surprised to learn that the interpretation of data is not always straightforward because data are subject to interpretation; yet, it is precisely because of this circumstance that practice is needed in interpreting research results as they actually appear in journals.

A statistical guide at the beginning of each exercise provides highlights that help in the interpretation of the associated excerpt. The guides are not comprehensive because it is assumed that students using this book are enrolled in statistics courses in which theoretical and computational concepts are covered in detail in their textbooks. Thus, the guides should be thought of as reminders of basic points to be considered when attempting the exercises.

Finally, students will discover occasional inconsistencies between what is recommended by their textbook authors and the analysis and reporting techniques employed by the authors of the excerpts. Variations are permitted by journal editors, and the excerpts in this book will help students prepare for reading published research articles that are not always "textbook perfect." When taking tests in class, however, students should follow the recommendations made by their textbook authors or by their instructors.

About the Third Edition

The first two editions of this book have been very popular. Hence, a light update seemed more appropriate than a major overhaul. Of the more than 42 exercises in the Second Edition, more than a dozen were replaced in order to keep the book timely. The exercises that are new to the Third Edition are: 1, 3, 5, 7, 13, 14, 21, 22, 24, 25, 29, 32, 38, and 40.

Acknowledgments

I am grateful to Richard Rasor, professor emeritus of American River College, and Robert Morman and Deborah Oh of California State University, Los Angeles, for the many suggestions they made for improving this book.

Zealure C. Holcomb

Exercise 1   Exposure of Youth to Violence
Percentage: I

Statistical Guide

Percent means "per one hundred." For example, if there are 1,000 residents in a town and 60% are Republicans, then, on the average, 60 out of each 100 are Republicans. To determine the total number of Republicans, multiply 1,000 by .60, which yields 600.

To calculate a percentage, divide the part by the whole and multiply by 100. For example, if 8 of the 234 seniors in a high school reported having tried cocaine, then 3.4% reported cocaine use (8/234 = .0342 x 100 = 3.42 = 3.42% or 3%).

Excerpt from the Research Article[1]

Participants consisted of 178 African American youth between the ages of 14 to 19 years… residing within the Detroit metropolitan area. The residency of participants spanned 26 zip codes within the metropolitan area. There were 96 females…and 82 males.

African American youth were recruited from four private high schools and three community youth centers.… It was not possible to survey all classrooms, and therefore 9th- and 11th-grade classrooms were surveyed at each school. This assured that both younger and older adolescents were included in the sample. There were no age or grade restrictions during community center recruitment.

All students in designated classrooms were eligible to participate in the study; however, only those [students] providing signed parental consent forms were administered surveys.… At community centers, youth were informed they could participate in a special activity if they obtained parental permission. Surveys were completed at the youth center. In all cases, participation was voluntary.

Statistics indicate that 85% of the youth had been victims of some type of violence (violence experienced), whereas 91% of adolescents had witnessed some form of violence (violence witnessed).

Table 1   *Percentages of violence exposure by gender*

Incidence of Violence	Victim		Witness	
	Male	Female	Male	Female
Being chased	62	23	80	59
Hit by family	53	55	51	74
Hit by nonfamily	49	32	64	57
Beaten or mugged	19	5	55	32
Sexually assaulted	2	20	10	8
Attacked with a knife	23	9	23	9
Seriously wounded	23	6	53	35
Shot or shot at	42	11	62	38
Shot at/shot someone	20	5	—	—
Witnessed a suicide	—	—	3	3
Witnessed a murder	—	—	14	11

---

[1] Source: Myers, M. A., & Thompson, V. L. S. (2000). The impact of violence exposure on African American youth in context. *Youth & Society, 32*, 253–267. Copyright © 2000 by Sage Publications, Inc. Reprinted with permission.

Questions for Exercise 1

Part A: Factual Questions

1. What percentage of the participants was male?

2. What percentage of the female participants witnessed someone being chased?

3. Did a larger percentage of males *or* a larger percentage of females witness someone being seriously wounded?

4. Expressed as a percentage, what was the difference between males witnessing a suicide and females witnessing a suicide?

5. How many of the females (not percentage) were victims of sexual assault?

6. How many males (not percentage) were seriously wounded (victims)?

7. How many more males than females were the victims of beatings/muggings? (Report your answer to a whole number.)

8. The males were most often the victims of what type of violence (violence experienced)?

9. The females were the least often witnesses to what type of violence (violence witnesses)?

Part B: Questions for Discussion

10. If you sum the percentages under the column labeled Victim/Male, you get considerably more than 100%. Does this make sense? How is this possible?

11. Would you feel comfortable in generalizing the results reported here to African American youth in public schools in the same metropolitan area? Explain.

12. Do you think that making participation voluntary was a good idea? Could this decision affect the validity of the results?

13. Do you think that when selecting students from the private high schools, selecting only students in the 9th and 11th grades was a good idea? Explain.

14. Do the results of this study surprise you? Why? Why not?

Exercise 2   Food Sources of Homeless Adults

Percentage: II

Statistical Guide

To review percentages, see the statistical guide for Exercise 1.

Excerpt from the Research Article[1]

A survey of 529 homeless men and women was conducted in Los Angeles County at 19 sites where homeless persons tended to congregate. These sites included emergency shelters, a parking lot, parks, a shopping mall, a large beach area, soup kitchens, food distribution centers, and job referral-social service assistance centers.

For the present analysis, seven women who reported being pregnant were eliminated from the sample. In addition, those who did not reveal their age or monthly income or did not allow the anthropometric measures to be collected were also deleted. This elimination process left a total of 457 participants (344 men, 113 women) ranging in age from 16 to 78 years. The ethnic composition of the sample was 64 percent white, 25 percent African American, 6 percent Hispanic, 5 percent Native American, and less than 1 percent Asian or other.

The table lists their food sources during the past week. The most common free sources were soup kitchens, missions or shelters, or friends. Only 6 percent of the sample did not use any free food sources. The most common other food sources used during the past week included a restaurant or a store.

Food sources of 457 homeless adults in Los Angeles, past week

Food source	Percent
Free sources:[1]	
Soup line	53
Mission, shelter	51
Friend	47
Food pantry	34
Garbage can	18
Relative	13
Number of free sources:	
None	6
One	30
Two	28
Three	21
Four	11
Five to six	4
Other sources:[1]	
Restaurant	52
Vending machine	20
Store	67

[1]Numbers total more than 100 percent because respondents could have used more than one source during the past week.

---

[1] Source: Gelberg, L., Stein, J., & Neumann, C. G. (1995). Determinants of undernutrition among homeless adults. *Public Health Reports, 110,* 448–454.

Questions for Exercise 2

Part A: Factual Questions

1. What was the single most prevalent source of free food?

2. What percentage of the 457 homeless adults obtained food from garbage cans?

3. How many of the 457 homeless adults obtained food from garbage cans?

4. How many of the 457 homeless adults obtained food from restaurants?

5. How many of the 457 homeless adults were Native American?

6. Of the 457 homeless adults studied, 114 of them were African American (25% of 457 = 114). If there had been only 100 homeless adults in the total sample, how many of them probably would have been African American?

7. The table is divided into three sections: Free sources, Number of free sources, and Other sources. The percentages sum to 100% in which one of the sections?

8. Men constituted what percentage of the 457 homeless adults? (Round your answer to one decimal place.)

9. Women constituted what percentage of the 457 homeless adults? (Round your answer to one decimal place.)

10. What is the sum of your answers to questions 8 and 9?

11. What percentage of the 529 homeless adults was eliminated/deleted from the sample? (Round your answer to one decimal place.)

Part B: Questions for Discussion

12. In your opinion, is the sample probably representative of all homeless adults in Los Angeles County? Explain.

13. The authors did not give the table a number. (Note that there was only one table in the article from which the excerpt was drawn.) Do you think the authors should have given the table a number? Explain.

14. In the last paragraph of the excerpt, the authors describe only the highlights of the entries in the table. Do you think it is acceptable to describe only the highlights, or do you think the authors should have discussed each entry in the table?

15. In the table, the "Free sources" are listed in descending order from the highest percentage to the lowest percentage. The "Number of free sources" and "Other sources" are *not* listed in descending order. Do you think they should be? Explain.

16. Do the results of this study surprise you? Why? Why not?

Exercise 3   Opiate Addicts Seeking Treatment
Frequency Distribution with Percentages

Statistical Guide

A frequency distribution helps to organize and summarize data. It is a statistical table that shows the number of cases (i.e., frequency of cases that obtained each score). Typically, the scores are listed in order (such as from high to low) in the first column, and the numbers of cases (*n*) are listed in the second column. The percentage of cases associated with each score is often also provided.

Excerpt from the Research Article[1]

[Little] is known about heroin and opium use in the Iranian population....Opium was known to the ancient Persians and has been traditionally used for recreation, for relieving pain, and also for treating mental disorders.

The data were gathered from 306 consecutive addicts who sought treatment at the Shiraz Self-Identified Addicts Center from July to September [in a recent year]. A semistructured interview was carried out with the subjects and one of their first-degree relatives.

The data were gathered from 306 subjects whose mean (a type of average) age was 37.0 years.... Subjects were 97.7% men of whom the majority (73.9%) were married and 24.2% single.

Duration of substance currently used is shown in Table 1. About 36% of the addicts reported that they had been using the current substance for more than a decade. Only 2.3% reported one year or less as the duration of taking the substance.

Table 1 gives the frequency distribution for the longest duration of abstinence. The majority (76.8%) gave a history of abstinence. Only 1.6% reported five years or more as the longest duration of abstinence, while 18.3% indicated one to three months as the longest duration. Overall, 62.1% relapsed before one year of abstinence was completed.

---

[1] Source: Ahmadi, J., & Ghanizadeh, A. (2000). Motivations for use of opiates among addicts seeking treatment in Shiraz. *Psychological Reports*, *87*, 1158–1164. Copyright © by Psychological Reports. Reprinted with permission.

Table 1 *Frequency distribution of addicts by duration of current opiate use and longest duration of abstinence reported (n = 306)*

Current use, yr.	n	%	Duration of abstinence	n	%
< 1	7	2.3	No abstinence	63	20.6
2	19	6.2	< 1 week	18	5.9
3	18	5.9	1 to 4 wk.	28	12.4
4	19	6.2	1 to 3 mo.	56	18.3
5	31	10.1	3 to 6 mo.	44	14.4
6	27	8.8	6 to 12 mo.	34	11.1
7	27	8.8	1 to 3 yr.	29	9.5
8	18	5.9	3 to 5 yr.	6	2.0
9	6	2.0	> 5 yr.	5	1.6
10	24	7.8	No answer	13	4.2
11–15	26	8.4			
16–20	57	18.6			
> 20	27	8.8			

Questions for Exercise 3

Part A: Factual Questions

1. At the time of the study, how many of the subjects had been using opiates for 10 years?

2. At the time of the study, what percentage of the subjects had been using opiates for 10 years?

3. Which current years-of-use category has the largest percentage of subjects?

4. Which duration-of-abstinence category has the largest number of subjects in it?

5. Does the category that you named in question 4 also have the largest percentage for duration of abstinence?

6. At the time of the study, how many of the addicts had been abstinent for less than one week?

7. Does the last column of percentages in the table sum to 100%?

8. Check the researchers' calculation of one of the percentages by dividing the number who reported no abstinence by the total number in the sample and multiplying by 100. To *two* decimal places, what answer do you get? (Note that the researchers reported the percentages to *one* decimal place.)

9. If you put all the names of the addicts in a hat, mix the names thoroughly, and pull one name out at random, the person you selected would most likely be in which duration-of-absence category?

Part B: Questions for Discussion

10. In the table, the researchers list the scores (years) from low (at the top) to high (at the bottom). If you have a textbook that covers this topic, does your textbook author recommend this arrangement? Explain.

11. Most of the scores for current years of use (in the first column) are for one-year periods. Yet some are for larger periods (such as 11–15), and one is for a smaller period (< 1). If you have a textbook that covers this topic, does your textbook author recommend that all of the score intervals be of the same size? Explain.

12. For duration-of-abstinence scores, the score with the lowest percentage is > 5 years. Do you think this indicates that abstaining from opiates over a number of years is very difficult? Explain.

13. Statement: "The percentage in the 16–20 current-years-of-use category is much larger than those in the 9 current-years-of-use category." While this is a correct statement, is there something potentially misleading about it? Explain.

Exercise 4   Changes in the Incidence of Asthma
Alternatives to Percentages and Proportions

Statistical Guide

A *percentage* is the *rate per 100*. To review percentages, see the statistical guide for Exercise 1. When a characteristic is very rare, percentages can be awkward to read and interpret. For example, 0.023% of the males in Canada committed suicide in a recent year. This is read as "twenty-three thousandths of one percent," which is quite a mouthful.

To convert a very small percentage to a different rate, use the following multipliers.

If you want to convert a percentage to this rate:	Multiply the percentage by this number: (*Multiplier*)	Example (Canadian male suicide rate)
Per 1,000	10	0.023% x 10 = 0.23 per 1,000
Per 10,000	100	0.023% x 100 = 2.3 per 10,000
Per 100,000	1,000	0.023% x 1,000 = 23 per 100,000
Per 1,000,000	10,000	0.023% x 10,000 = 230 per 1,000,000

To convert some other rate to a percentage, use the following divisors.

If you have one of the following rates and want to convert it to a percentage:	Divide the rate by this number: (*Divisor*)	Example (Canadian male suicide rate)
Per 1,000	10	0.23 per 1,000 becomes: 0.23 ÷ 10 = 0.023%
Per 10,000	100	2.3 per 10,000 becomes: 2.3 ÷ 100 = 0.023%
Per 100,000	1,000	23 per 100,000 becomes: 23 ÷ 1,000 = 0.023%
Per 1,000,000	10,000	230 per 1,000,000 becomes: 230 ÷ 10,000 = 0.023%

Excerpt from the Research Article[1]

Figures 1 and 2 show the trends in the prevalence of asthma and the death rate for asthma.

---

[1] Source of the data: Centers for Disease Control and Prevention (1995). Asthma—United States, 1982–1992. *MMWR Morbid Mortal Weekly Report, 43*, 953–955. Figures reproduced from Brown, C. M., Etzel, R. A., & Anderson, H. A. (1997). Asthma: The states' challenge. *Public Health Reports, 112*, 198–205.

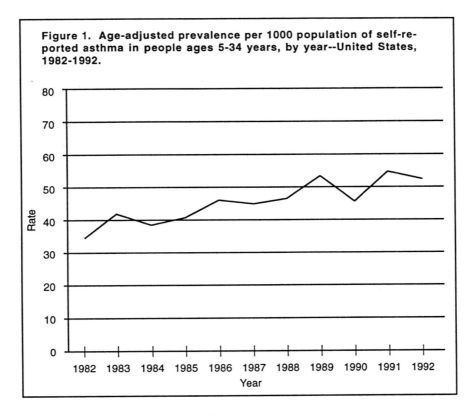

Figure 1. Age-adjusted prevalence per 1000 population of self-reported asthma in people ages 5-34 years, by year--United States, 1982-1992.

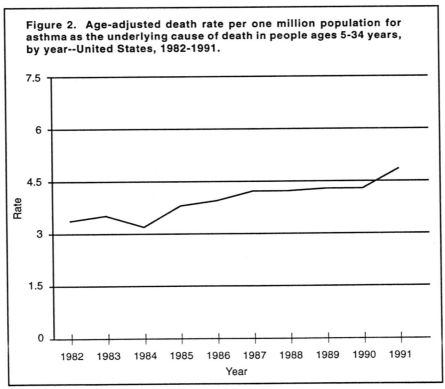

Figure 2. Age-adjusted death rate per one million population for asthma as the underlying cause of death in people ages 5-34 years, by year--United States, 1982-1991.

Questions for Exercise 4

Note: All questions refer to people who are 5 to 34 years of age.

Part A: Factual Questions

1. How many people per 1,000 reported having asthma in 1985?

2. What percentage of people in 1985 reported having asthma?

3. In 1991, about 54 in 1,000 reported having asthma. What percentage of the people reported having asthma in 1991?

4. According to Figure 2, about how many people in 1,000,000 died of asthma in 1991?

5. In 1990, about 4.4 people in 1,000,000 died with asthma as the underlying cause. What is the corresponding percentage?

6. Write your answer to question 5 in words without using numerals.

7. Suppose 0.15% of people died of another disease in a given year. How many people per 1,000 died of it?

8. For the percentage in question 7, how many people per 10,000 died of it?

Part B: Questions for Discussion

9. Speculate on why the authors did not use the same rate in both figures.

10. The rate of self-reported asthma dropped from 1991 to 1992 (see Figure 1). Should this be cause for optimism?

11. Describe in your own words the overall trend in the data presented in Figure 2.

12. The authors placed the figure numbers and titles (i.e., captions) above the figures. Check in your textbook and any style manuals for your field to see if this is the recommended placement. Write your findings here.

Exercise 5   Drug Usage Trends
Time Series Line Graph

Statistical Guide

Figure 1 shows two time series line graphs. A line graph shows changes over time. The time intervals (in this case, ages) are usually shown on the horizontal axis (i.e., the *x*-axis).

Excerpt from the Research Article[1]

These line graphs are based on data from a large national survey conducted at 47 sites serving high-risk youth.

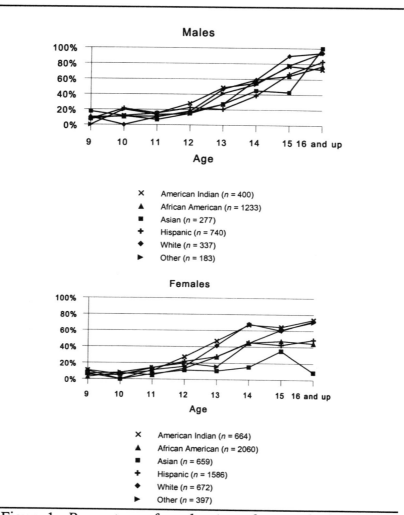

Figure 1   *Percentage of youth using tobacco, alcohol, or marijuana during the past 30 days.*

---

[1] Source: Chipungu, S. S. et al. (2000). Prevention programming for African American youth: A review of strategies in CSAP's national cross-site evaluation of high-risk youth programs. *Journal of Black Psychology, 26,* 360–385. Copyright © 2000 by Sage Publications, Inc. Reprinted with permission.

Questions for Exercise 5

Part A: Factual Questions

1. The vertical axis (in this case, the one with the percentages) is known as the
   A. *x*-axis.   B. *y*-axis.

2. Is the number of American Indian males the same as the number of African American males in this study? Explain.

3. At age "16 and up," which group of females had the smallest percentage using tobacco, alcohol, or marijuana?

4. About what percentage of Asian males were using the substances at age 9?

5. Consider both males and females. Which group showed the largest percentage decrease from age 15 to age "16 and up"?

6. Between which two ages did the Asian males show the sharpest percentage increase?

7. Notice that this excerpt contains a *figure* rather than a *table*. (See Exercise 1 on page 1 for an example of a table.) What is the difference between a *statistical figure* and a *statistical table*?

Part B: Questions for Discussion

8. Even though there are more Hispanic females than White females, both groups of females are plotted on the same line graph. Does this make sense? Explain.

9. From line graphs such as the one in this exercise, readers cannot always obtain the exact percentages for each group. Is this a major problem? Explain.

10. In many textbooks, the symbol "n" is italicized (as *n*). It is not italicized in this report. Is this a problem?

11. Each of the groups shown here is made up of various subgroups. For example, there are many different tribal groups that constitute American Indians. In your opinion, should researchers strive to use smaller subgroups in their analysis of data?

12. If you were conducting a study on the same topic, would you group tobacco, alcohol, and marijuana into the same line graph, or would you have separate ones for each substance? Explain.

Exercise 6 Television Viewing, Books, and Reading
Histogram

Statistical Guide

A *histogram* has vertical bars. The scores on a continuous variable are placed on the horizontal axis (i.e., *x*-axis). For example, age is a continuous variable because there are no gaps between the ages of 0 and 1, between the ages of 1 and 2, and so on. The vertical axis (i.e., *y*-axis) shows the frequency, percentage, or rate of occurrence.

An *outlier* is an observation that is far from the other observations. For example, Figure 3 clearly has outliers.

The *mode* in a histogram is the most frequently occurring score. In the figures below, the scores are "numbers of books," "viewing time in hours," and "amount of reading measured as the number of times in a two-week period."

Excerpt from the Research Article[1]

Thirty preschool children and their primary caregivers participated. Participants were recruited from Chapel Hill and Durham, North Carolina, public housing communities and Head Start programs.

We assessed the quality of the home environment with items inquiring about the number of children's books in the home, frequency of parent-child joint reading, and parental instruction.

We assessed television viewing with three items, which asked parents to estimate their children's viewing time separately for weekdays, Saturdays, and Sundays; these items were later combined to yield a single index of viewing time per week.

The sample frequencies for the variables reflecting the educational quality of the home environment are displayed in Figures 1–3.

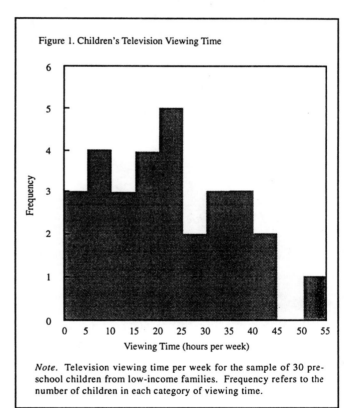

Figure 1. Children's Television Viewing Time

Note. Television viewing time per week for the sample of 30 preschool children from low-income families. Frequency refers to the number of children in each category of viewing time.

---

[1] Source: Clarke, A. T., & Kurtz-Costes, B. (May/June 1997). Television viewing, educational quality of the home environment, and school readiness. *The Journal of Educational Research*, Vol. *90*, No. 5, 279–285. Reprinted with permission from the Helen Dwight Reid Educational Foundation. Published by Heldref Publications, Washington, DC 20036-1802. Copyright © 1997.

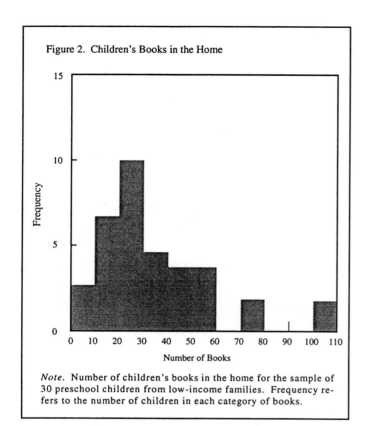

Figure 2. Children's Books in the Home

Note. Number of children's books in the home for the sample of 30 preschool children from low-income families. Frequency refers to the number of children in each category of books.

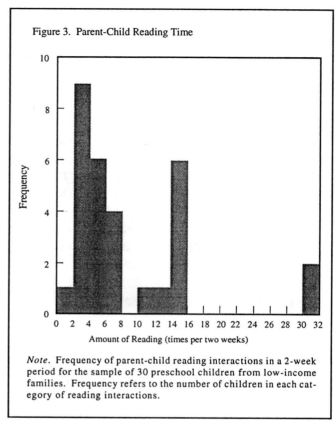

Figure 3. Parent-Child Reading Time

Note. Frequency of parent-child reading interactions in a 2-week period for the sample of 30 preschool children from low-income families. Frequency refers to the number of children in each category of reading interactions.

Questions for Exercise 6

Part A: Factual Questions

1. How many children watched between 40 and 45 hours of television per week?

2. Which number of hours of television viewing (e.g., 0 to 5, 5 to 10, etc.) had the largest number of children?

3. How many children's homes had 20 to 30 books?

4. How many children were engaged in parent-child reading interactions (during the two-week period) from 14 to 16 times?

5. What *percentage* of the children watched television between 20 and 30 hours per week? (If you do not recall how to calculate percentages from frequencies, see the Statistical Guide in Exercise 1.)

6. In Figure 1, what is the mode? (Hint: In this case, the mode is a range of scores.)

7. In Figure 2, what is the mode?

8. How many children are clearly outliers in Figure 3?

9. Do the bars in any of the histograms form a bell-shaped (i.e., normal) distribution?

10. Some authors of statistics textbooks recommend that the scores be grouped into about 10 to 20 *intervals* for presentation in a histogram. (Note: In Figure 1, the authors have *not* presented a bar for each number of hours but rather for *intervals* of hours such as 50 to 55.) Do all three histograms in the excerpt conform to the textbook authors' recommendation?

Part B: Questions for Discussion

11. Speculate on whether the first bar on the left in Figure 1 stands for 0 to 4 hours *or* 0 to 5 hours. (Note that if the first bar stands for 0 to 5 hours, the next one should stand for 6 to 10 hours.)

12. Textbook authors usually recommend that each histogram be given a number and a title, as was done in the histograms in this exercise. Speculate on why the authors make this recommendation.

13. The authors were interested in getting a single, overall estimate of each child's television viewing time per week. To do so, they asked about viewing time separately for weekdays, Saturdays, and Sundays. In your opinion, was it a good idea to ask about television viewing with three questions instead of only one that asks for total viewing time per week? Explain.

Exercise 7 Monthly Suicide Rates
Line Graph with Rate per 100,000

Statistical Guide

A line graph can show seasonal changes in behavior. They are especially interesting when they are averaged over a period of several years, which washes out idiosyncratic results that might occur, for example, in one particularly wet winter season.

The line graphs in this exercise show rates per 100,000. To review this concept, see the statistical guide for Exercise 4.

Excerpt from the Research Article[1]

This study attempts to examine the existence of seasonality in suicide rates and the relationship between weather and suicide rate.

Monthly data for suicide...from 1980 to 1994 were obtained from the Hong Kong Census and Statistics Department.

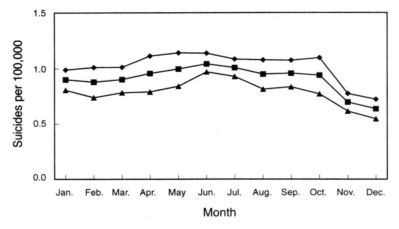

Fig. 1. Average total and sex-specific monthly suicide rates, 1980-1994.
(■ Total, ◆ Male, ▲ Female)

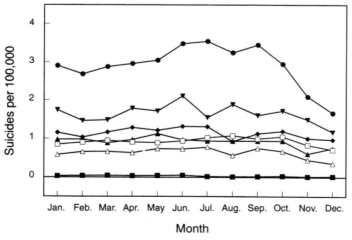

Fig. 2. Average age of specific monthly suicide rates, 1980-1994.
(■ < 16 yr., △ 16-24 yr., ▲ 25-34 yr., □ 35-44 yr., ◆ 45-54 yr., ▼ 55-64 yr., ● > 65 yr.)

[1] Source: Yan, Y. Y. (2000). Geophysical variables and behavior: LXXXXIX. The influence of weather on suicide in Hong Kong. *Perceptual and Motor Skills*. *91*, 571–577. Copyright © 2000 by Perceptual and Motor Skills. Reprinted with permission.

Questions for Exercise 7

Part A: Factual Questions

1. Is the average total suicide rate in January greater than *or* less than 1 in 100,000?

2. Are men *or* women consistently above the average total across all months?

3. Examine the average suicide rate for males in January in Figure 1. If these data applied to a city with 500,000 males, about *how many* should you expect to commit suicide in January?

4. Examine the average suicide rate for males in January in Figure 1. If these data applied to a city with 1,000,000 males, about *how many* should you expect to commit suicide in January?

5. You should expect what *percentage* of males in Hong Kong to commit suicide in January? (See the statistical guide for Exercise 4 if you do not know how to calculate the answer.)

6. Which age group consistently has the highest suicide rate across the months?

7. For people under 16 years of age in Hong Kong, what *percentage* should you expect to commit suicide in December?

8. Examine the average suicide rate for those older than 65 in April. If these data applied to a city with 700,000 people older than 65, about *how many* of them should you expect to commit suicide?

Part B: Questions for Discussion

9. That data in the figures are for a 15-year period. In your opinion, is this a sufficiently long period to identify reliable trends? Is it much better than examining only one year? Explain.

10. Do you consider any of the trends striking or noteworthy? Explain.

11. The data in the figures could have been presented in a table. Part of such a table is shown immediately below. Do you prefer to have the data presented in a line graph or in a table? Explain.

Table for Question 11 *Average suicide rates per 100,000*[1]

Month	Men	Total	Women
January	1.0	0.9	0.8
February	1.1	0.8	0.7
March	1.1	0.9	0.8
and so on.			

[1] Values are rough approximations from Figures 1 and 2.

12. The title of the research article from which these data are drawn refers to "the influence of weather on suicide." In your opinion, do these data strongly support the hypothesis that weather influences suicide rates? Could there be other seasonal explanations? Explain.

Exercise 8   Norms for a Spelling Test
Cumulative Percentage and Percentile Rank

Statistical Guide

A cumulative percentage indicates the percentage of examinees that scored at and below a given score. A cumulative percentage is also known as a percentile rank. For example, if 40 percent of a group had scores equal to or lower than an examinee's score, then that examinee has a percentile rank of 40. Note that percentile ranks are usually rounded to whole numbers when reported to examinees.

Test makers often try out a test with a large group of examinees (known as the norm group) and then build a table such as the one in the excerpt (known as a norms table). Those who subsequently take the test can convert their raw score (number right) to a percentile rank using the table.

Excerpt from the Research Article[1]

This test is composed of 76 words with one or more letters missing. A short line indicates where the letter(s) should be inserted (e.g., exper__ment). Words were selected from lists of the words most frequently misspelled by college students. Brief hints are provided in ambiguous or potentially difficult situations (e.g., capt__n/military rank). The test is timed at 10 minutes.

A sample of 316 undergraduate university students was tested on the Spelling Component Test. Results from this administration were combined with the 386 subjects tested [earlier] to furnish a normative sample of 702 university students (407 females and 295 males).

The normative distribution of scores, for the full sample and separated by gender, is given in Table 3, in terms of cumulative percentages. **[See Table 3 on the next page.]**

Questions for Exercise 8

Part A: Factual Questions

1. What percentage of the females had scores at and below 27–28?

2. What percentage of the males had scores at and below 27–28?

3. Compare your answers to questions 1 and 2. Based on these two answers, does it appear that females *or* males performed better on the spelling test? Explain.

---

[1] Source: Coren, S. (1989). The Spelling Component Test: Psychometric evaluation and norms. *Educational and Psychological Measurement, 49,* 961–971. Copyright © 1989 by Educational and Psychological Measurement. Reprinted by permission.

Table 3 *Norms for the Spelling Component Test based on*
702 university undergraduate students

Raw Score	Cumulative Percentage		
	Females (N = 407)	Males (N = 295)	Total (N = 702)
69–70	99.4	99.6	99.5
67–68	98.6	99.6	99.0
65–66	97.1	99.6	98.2
63–64	94.8	97.1	95.9
61–62	92.5	95.5	93.9
59–60	88.0	93.0	90.5
57–58	83.0	87.6	85.2
55–56	76.1	84.7	79.8
53–54	70.9	79.3	74.7
51–52	67.4	76.0	71.4
49–50	63.4	72.3	67.5
47–48	59.0	68.2	63.2
45–46	53.0	61.2	56.7
43–44	45.0	53.3	48.8
41–42	36.9	49.2	42.5
39–40	31.7	43.8	36.8
37–38	25.6	38.2	31.0
35–36	20.5	30.2	25.0
33–34	16.4	24.4	20.4
31–32	11.2	17.8	14.3
29–30	8.1	14.5	11.2
27–28	6.1	11.2	7.9
25–26	4.0	8.7	5.7
23–24	2.0	6.2	3.6
21–22	1.4	4.1	2.5
19–20	0.9	3.7	2.0
17–18	0.3	1.2	1.0
15–16	0.3	0.4	0.3

4. Based on the total sample, what is the percentile rank for a person with a raw score of 34?

5. Based on the norms for males, what is the percentile rank for a male with a raw score of 34?

6. Based on the total sample, a percentile rank of about 85 corresponds to which raw scores?

7. What percentage of the females had scores of 38 or less?

8. What percentage of the males had scores of 38 or less?

9. If the test were being used for college admissions, would a female who had a raw score of 38 be better off if her percentile rank being used for admissions purposes was derived from the female norms *or* the norms for the total sample (both males and females)? Explain.

10. If the test were being used for college admissions, would a male who had a raw score of 38 be better off if his percentile rank being used for admissions purposes was derived from the male norms *or* the norms for the total sample (both males and females)? Explain.

Part B: Questions for Discussion

11. The researcher states that there were 76 words on the test, but the highest score in Table 3 is 70. Speculate on the reason for this apparent discrepancy.

12. Suppose your professor administered the spelling test to you and offered to give you either your raw score only *or* your percentile rank based on your gender only. Which would you choose? Why?

13. The norms for the total sample are based on the responses of more women than men. Is this a problem? Explain.

Exercise 9   Penalties for Providing Alcohol to Minors

Mean, Median, and Range

Statistical Guide

The mean is the average that is the balance point in a distribution. It is calculated by summing all the scores and dividing by the number of scores. The mean is pulled toward extreme scores in an unbalanced distribution (i.e., a skewed distribution with extreme scores on one side without extreme scores on the other side to balance it).

The median is the average that indicates the value below which half of cases lie. For example, if the median for a group is 10.0, then 50 percent of the cases lie below 10.0.

The range is often defined as the difference between the highest score (maximum) and the lowest score (minimum) in a distribution, which is the definition you should use in this exercise. (Some statisticians add "1" to the difference.)

Excerpt from the Research Article[1]

We selected four states for intensive study: Kentucky, Michigan, Montana, and Oregon. Selection was based on the availability of detailed enforcement data for both ABC [Alcoholic Beverage Control] agencies and local police departments, and on the diversity of the states in terms of their alcohol control systems. However, the states were not selected randomly, and generalizations to other states and localities must therefore be made cautiously.

Rates of youth drinking arrests across the 295 counties are highly skewed.... Low levels of enforcement actions on underage drinking are clear in the number of actions per year in the average county (Table 2). The median county has no liquor license suspensions or revocations, 1 ABC action against an alcohol outlet, 8 arrests for possession, and a total of 26 liquor law arrests of 16–20-year-olds.

Table 2   *Descriptive statistics on enforcement of drinking age: Counts for 295 counties in Kentucky, Michigan, Montana, and Oregon*

Statistic	Arrests for law violations		ABC actions against persons for selling to persons younger than 21	ABC suspensions of licenses for supplying to persons younger than 21	ABC revocations for selling to persons younger than 21
	Total arrests (ages 16–20)	Possession (ages 16–20)			
Minimum	0	0	0	0	0
Maximum	3,905	1,223	288	29	12
Median	26	7.7	1.0	0	0
Mean	220	52	6.0	.59	.25

Note: ABC = Alcoholic Beverage Control

We found that rates of enforcement of the legal minimum drinking age are very low, particularly in terms of actions taken against those who sell or provide alcohol to underage youth. Many counties give no attention to drinking age enforcement at all. Twelve percent of the counties examined had no arrests of youth younger than age 21 for illegal possession of alcoholic beverages

---

[1] Source: Wagenaar, A. C., & Wolfson, M. (1995). Deterring sales and provision of alcohol to minors: A study of enforcement in 295 counties in four states. *Public Health Reports, 110,* 419–427.

across the entire 3-year period examined. When enforcement actions are taken, they typically are focused on the individual young drinker, rather than on the commercial outlet…that supplied the alcoholic beverages to youth.

Conversely, there are a small number of jurisdictions with very high rates of enforcement actions on underage drinking, suggesting that barriers to enforcement of this law are not insurmountable.

Questions for Exercise 9

Part A: Factual Questions

1. What is the largest number of arrests of 16–20-year-olds for possession in any county?

2. What is the range of number of arrests for possession?

3. What is the largest number of ABC revocations in any one county?

4. What is the range for ABC revocations?

5. The excerpt states that the drinking arrests data are "highly skewed." When this is the case, which of the following should you expect?

    A. The mean and median will be very similar in value.
    B. The mean and median will be clearly different in value.

6. Do the statistics in Table 2 confirm your answer to question 5? Explain.

7. The data for arrests has what type of skew?

    A. Skewed to the right (i.e., positive skew) with a small number of counties with a large number of arrests.
    B. Skewed to the left (i.e., negative skew) with a small number of counties with a small number of arrests.

8. What percentage of the counties had fewer than 1 ABC action against persons for selling to persons younger than 21?

9. What percentage of the counties had more than 26 total arrests of those aged 16–20 for alcohol law violations?

Part B: Questions for Discussion

10. In Table 2, the mean for the total number of arrests of 16–20-year-olds is 220. Is it possible to determine from this the percentage of counties that had less than 220 by knowing the mean? Explain.

11. Suppose someone who had not studied statistics examined Table 2 and asked you, "What is the average number of ABC actions against persons for selling to persons younger than 21?" How would you answer the question?

12. The authors state that "generalizations to other states and localities must therefore be made cautiously." Do you agree? Explain.

Exercise 10   Injection Drug Users and HIV

Median and Interquartile Range: I

Statistical Guide

To review the median, see the statistical guide for Exercise 9.

The interquartile range indicates the scores obtained by the middle 50 percent of the participants. Put another way, if all the scores are put in order from low to high and the bottom 25 percent and the top 25 percent are temporarily ignored, the range of the remaining scores is the interquartile range. This statistic is a measure of *variability* (also known as *spread* or *dispersion*) that is usually reported in conjunction with the median.

Excerpt from the Research Article[1]

This analysis was conducted as part of the World Health Organization (WHO) Multi-Centre Study of AIDS and Injecting Drug Use. Persons who had injected illicit drugs within the previous 2 months were recruited from drug abuse treatment programs and nontreatment settings (in most cities, through outreach and chain-referral sampling). It is estimated that at least 95% of the subjects who were asked to participate agreed to do so.

The questionnaire focused on drug use histories and on injection and sexual risk behavior in the 6 months prior to the interview.... After the interview, either a blood or saliva specimen was obtained for HIV testing.

A specific series of questions was used to ascertain subjects' deliberate behavioral changes in response to concerns about AIDS. Subjects were asked, "Since you first heard about AIDS, have you done anything to avoid getting AIDS?" Those who responded "yes," that they had changed their behavior, were then asked, "What have you done?"

In each of the four cities [in Table 1], changes in drug injection behavior were reported more often than changes in sexual behavior. The most commonly reported risk reduction was the "stopped/reduced sharing" of injection equipment. The most commonly reported sexual risk reductions were an increased use of condoms, greater selectivity in choosing sexual partners, and a reduced number of sexual partners.

Table 1   *Demographic characteristics, reported AIDS risk reduction, and HIV seroprevalence among injection drug users by city*

	Median Age (Interquartile Range)	Male %	Median Education (Interquartile Range)	Median Years Injecting (Interquartile Range)	Reported Risk Reduction %	HIV Positive %
Bangkok (*n* = 590)	30 (25–33)	95	7 (4–10)	8 (3.5–13)	92	34
Glasgow (*n* = 452)	23 (21–26)	70	11 (10–11)	6 (4–8)	83	2
New York (*n* = 829)	37 (31–41)	76	11 (10–12)	19 (10–24)	79	48
Rio de Janeiro (*n* = 128)	30 (25–33)	83	12 (9–14.5)	10 (4.5–14)	58	35

---

[1] Source: Des Jarlais, D. C., Friedmann, P., Hagan, H., & Friedman, S. R. (1996). The protective effect of AIDS-related behavioral change among injection drug users: A cross-national study. *American Journal of Public Health, 86,* 1780–1785. Copyright © by the American Public Health Association.

Questions for Exercise 10

Part A: Factual Questions

1. On the average, the participants in which city were oldest? Explain.

2. In Bangkok, what percentage of the participants had less than 7 years of education?

3. In New York, what percentage of the participants had more than 11 years of education?

4. On the average, the participants in which city had the fewest years of injecting?

5. The participants in which city had the greatest variability in their ages? Explain.

6. The participants in which city had the greatest dispersion in their number of years of education?

7. In New York, what percentage of the participants had been injecting for 10 to 24 years?

8. In New York, what percentage of the participants had been injecting for less than 10 years?

9. In New York, *how many* of the 829 participants had been injecting for more than 24 years?

10. In Rio de Janeiro, what percentage of the participants had been injecting from 4.5 to 14 years?

11. In Rio de Janeiro, *how many* of the 128 participants had been injecting for more than 14 years?

12. The participants in which city had the least variability in the number of years of injecting?

Part B: Questions for Discussion

13. Many textbook authors suggest that the interquartile range be reported as a single value obtained by subtracting the two values shown in the excerpt. For example, in the excerpt, the interquartile range is shown as "25–33" for the age of Bangkok's participants. Using the textbook authors' suggestion, it would be reported as "8" (i.e., 33 – 25 = 8). Which method of reporting do you prefer? Why?

14. The median age of the participants in New York is substantially higher than in the other cities. Could this help explain some of the differences in median years of injecting across cities? Explain.

15. The city with the lowest percentage of HIV+ participants had the lowest median age. Does this make sense? Explain.

16. When considering data on HIV risk reduction behavior, would you be interested in an analysis that considers men and women separately? Explain.

17. Speculate on what the authors mean by "chain-referral sampling."

Exercise 11   Adolescent Physical Abuse and Suicide

Median and Interquartile Range: II

Statistical Guide

To review the median and interquartile range, see Exercise 10.

To obtain the interquartile range, you must identify the 25th percentile (the score below which 25% of the participants scored) and the 75th percentile (the score below which 75% of the participants scored).

Note that in this excerpt the authors refer to *significant differences*. You will be learning about this topic in detail later in this book. At this point, *significant differences* may be thought of as *reliable differences*. Usually, significant differences are large enough to be of interest in the interpretation of results.

Excerpt from the Research Article[1]

The sample, representing a white, middle-class, suburban population, consisted of 99 physically abused adolescents from Nassau and Suffolk Counties.... [Adolescents who had suffered intrafamilial *sexual* abuse were not included in the sample of 99 abused adolescents.]

In the abuse sample, biological fathers were most frequently indicated as the perpetrators of adolescent abuse (73% of the cases). Mothers were indicated as perpetrators in 25% of the cases. Stepfathers were indicated as the perpetrators in 10% of the cases. In 10 families, more than one perpetrator was indicated.

For this study, a suicide attempt was defined as any intentional, self-inflicted injury accompanied by a statement of suicidal intent, or classic severe suicidal injuries such as a large ingestion of toxic substances (e.g., 20 pills or more at one time), self-inflicted deep wounds to wrist or throat, unsuccessful hanging, or gunshot wounds to the head or abdomen.

[The Family Adaptability and Cohesion Scales] is a 20-item, paper-and-pencil self-report measure... [For this scale,] cohesion is defined as "the emotional bonding that family members have toward each other"...and adaptability is defined as "the ability of a marital or family system to change its power structure, role relationships, and relationship rules in response to situational and developmental stress." [Higher scores indicate more cohesion and more adaptability.]

[Academic performance was measured with the Youth Self-Report. In the national norm group, the mean score for academic performance was 50.]

Table 3 presents comparative data on indicators of adolescent vulnerability leading to mental illness and suicidal behavior. Abused attempters were not significantly different from abused nonattempters in peer social support: No significant differences were found in either the number of close friends they cited or reported they enjoyed being with. However, they reported inadequacies in family support. Abused attempters perceived their mothers as being less caring (median = 16.5) than did abused nonattempters (median = 29)...and they perceived their families to be less cohesive (median = 20.5) than did abused nonattempters (median = 28).

---

[1] Source: Kaplan, S. J., Pelcovitz, D., Salzinger, S., Mandel, F., & Weiner, M. (1997). Adolescent physical abuse and suicide attempts. *Journal of the American Academy of Child and Adolescent Psychiatry, 36*, 799–808. Copyright © 1997 by Williams & Wilkins. Reprinted by permission.

Table 3 *Indices of vulnerability in abused attempters and abused nonattempters*

	Attempters		Nonattempters	
	Median	IQR	Median	IQR
Peers				
No. of close friends	3	0–4	4	1–8
No. of peers enjoy	2.5	0–3.5	4	1–7
Family				
Perception of mother's caring	16.5	13.0–27.5	29	20–32
Perception of father's caring	14.5	6.5–21.5	19.5	14–26
Family cohesion	20.5	17.5–26	28	23–25
Family adaptability	24	19.0–25.5	22	10–26
Academic performance				
YSR score	35	30–47	42	35–51

Note: IQR = interquartile range (25th to 75th percentile); YSR = Youth Self-Report

Questions for Exercise 11

Part A: Factual Questions

1. On the average, which group reported having more close friends?

2. On the average, which group reported having lower academic performance?

3. What percentage of the nonattempters reported having between 1 and 8 close friends?

4. What percentage of the nonattempters reported having more than 8 close friends?

5. What percentage of the attempters reported having more than 4 close friends?

6. What is the 75th percentile for number of close friends for nonattempters?

7. What is the 25th percentile for number of close friends for nonattempters?

8. Which group was more variable in the number of close friends? Explain.

9. Which group was less variable in family cohesion? Explain.

10. Did any of the middle 50% of attempters have a YSR score as high as the national mean of 50?

11. What is the average difference between the two groups in family cohesion scores?

12. What is the median difference between the two groups in perception of mother's caring?

Part B: Questions for Discussion

13. Speculate on why adolescents who had been sexually abused were not included in the study.

14. If you have a statistics textbook, examine the definition of the interquartile range. Are the interquartile ranges in this excerpt reported as suggested in your textbook? (Hint: For each variable in the excerpt, the IQR is reported as two score values.) Explain.

15. The authors state in the excerpt that the interviewers "were blind with respect to the subjects' abuse status." Speculate on what this means. Is it important? Why? Why not?

16. The authors chose to report the median and interquartile range instead of the mean and standard deviation. If you have a statistics textbook, examine it to determine under what circumstances this is recommended. Write your findings here.

Exercise 12 Nutrition of College Students and Nonstudents
Mean and Standard Deviation

Statistical Guide

The mean is the average that is the balance point in a distribution. The standard deviation is a yardstick for measuring differences among a group of participants. It is a special type of average deviation from the mean; the larger the differences between the scores of a group and their mean, the larger the standard deviation. For example, if Groups A and B both have a mean of 10.00, then they are the same on the average. However, if the standard deviation for Group A equals 3.00 and the standard deviation for Group B equals 2.00, we know that the difference between the scores for Group A and their mean are greater than for Group B. Put another way, we can say that Group A has more *spread* or *dispersion* in its scores. Thus, we can infer that Group A's scores are more spread out, dispersed, or different because the standard deviation indicates that they are more distant from their mean.

Excerpt from the Research Article[1]

As part of a research project on food choices, researchers in 9 of 10 participating states (Arizona, Idaho, Iowa, Kansas, Michigan, Nebraska, New York, Oregon, Wisconsin) purchased names and addresses in random order by zip code for each state and telephone numbers of 18- to 24-year-olds.... Names were chosen randomly from the lists, and all but one state verified age and asked for agreement to participate by telephoning prospective participants. Those who agreed to participate were mailed a questionnaire and a postage-paid return envelope. Participants' names were entered into a drawing for a $25 reward or a gift as an incentive to participate. Researchers in 7 of the 9 participating states conducted follow-up telephone calls and mailings to nonrespondents. The response rate ranged from 25% to 72% among the states.

Table 1 *Self-reported personal characteristics and habits of young adults by gender and student status**

	Women		Men	
	Students	Nonstudents	Students	Nonstudents
Variable	($n = 412$)	($n = 219$)	($n = 328$)	($n = 194$)
Mean age (years)**	20.7 ± 1.6	22.1 ± 1.7	21.1 ± 1.7	21.9 ± 1.8
Mean times per week breakfast is eaten	4.1 ± 2.4	3.7 ± 2.4	4.1 ± 2.3	3.4 ± 2.4
Mean times per week lunch is eaten	5.8 ± 1.5	5.7 ± 1.7	6.0 ± 1.4	5.9 ± 1.5
Mean number of alcoholic drinks per week	2.0 ± 4.5	2.1 ± 7.2	4.8 ± 8.4	4.3 ± 7.4
Importance of eating the most nutritious foods***	3.1 ± 0.7	2.9 ± 0.8	2.7 ± 0.8	2.6 ± 0.8
Ability to use nutrition labels***	3.1 ± 0.8	2.9 ± 0.8	2.8 ± 0.9	2.6 ± 0.9
Adequate money to buy food***	2.8 ± 0.7	2.9 ± 0.7	2.8 ± 0.8	2.9 ± 0.7

*Students were currently college students; nonstudents were not college graduates or current students
**Means ± standard deviations
***Means of a four-point scale from 1 = not important to 4 = very important or 1 = very inadequate to 4 = very adequate

[1] Source: Georgiou, C. C., Betts, N. M., Hoerr, S. L., Keim, K., Peters, P. K., Stewart, B., & Voichick, J. (1997). Among young adults, college students and graduates practiced more healthful habits and made more healthful food choices than did nonstudents. *Journal of the American Dietetic Association, 97*, 754–759. Copyright © by the American Dietetic Association. Reprinted by permission of the American Dietetic Association, Volume 97.

Questions for Exercise 12

Part A: Factual Questions

1. On the average, which group of women was older? Explain.

2. On the average, which of the four groups had a higher score on the importance of eating the most nutritious foods?

3. On the average, which of the four groups reported eating breakfast the least often?

4. What is the average difference in age between the women students and the women nonstudents?

5. For the men students, which of the following is higher?
   A. The average number of times breakfast is eaten.
   B. The average number of times lunch is eaten.

6. Which group of men had more variability in the number of alcoholic drinks per week? Explain.

7. The two groups of women are very similar with respect to the average number of alcoholic drinks they consume per week. Are they very similar in their variability? Explain.

8. Which group of women had less spread in the number of times lunch is eaten each week?

9. Which of the four groups had greater dispersion in terms of having adequate money to buy food?

Part B: Questions for Discussion

10. Based on your answer to question 7, would you expect to find more problem drinkers among the women students or the women nonstudents? Explain.

11. For comparing a group of students with a group of nonstudents, is it important to have two groups that are similar in age? Explain.

12. What is your opinion on asking students to rate the importance they place on eating nutritious foods? In your opinion, will this result in useful data? Explain.

13. What is your opinion on the incentive used to encourage participation? Would it have encouraged you to participate? Explain.

Exercise 13   Self-Reported Height and Weight
Mean, Standard Deviation, and 68% Rule

Statistical Guide

The standard deviation is a yardstick for measuring variability (i.e., differences among participants, subjects, respondents, or cases). Synonyms for *variability* are *spread* and *dispersion*. The larger the value of the standard deviation, the greater the variability.

In a normal distribution (i.e., a type of symmetrical, bell-shaped distribution), about 34% of the cases lie within one standard deviation unit of the mean. For example, if the mean equals 50.00 for a normal distribution and one standard deviation unit equals 10.00, then 68% of the cases lie within (i.e., plus and minus) 10 points of the mean (i.e., 68% lies between 40.00 and 60.00).

Excerpt from the Research Article[1]

Accuracy in height and weight characteristics allows accuracy in calculating placement in health categories. Typically, questionnaires on health-risk appraisal and insurance applications require participants to include their [own estimates]....

A convenience sample of 62 (n = 30 men, n = 32 women) Euro-American college students with a mean age of 19.9 yr. (SD = 2.9) volunteered for the study.

Subjects were asked to volunteer for a simple assessment of fitness behavior via questionnaire, which included items of height and weight, but were not told prior to completing the survey that height and weight would be measured [using a scale]. Heights and weights were measured to the nearest 0.125 in. (.0032 m) and .25 lb (.113 kg) and were converted to meters and kilograms, respectively.

Table 1   *Comparison of self-reported and measured height and weight for college men (n = 30) and women (n = 32)*

Condition		M	SD
Men	Self-reported height, m	1.79	0.07
	Measured height, m	1.78	0.06
Women	Self-reported height, m	1.62	0.07
	Measured height, m	1.62	0.07
Men	Self-reported weight, kg	79.7	13.39
	Measured weight, kg	79.2	14.44
Women	Self-reported weight, kg	61.2	9.06
	Measured weight, kg	63.1	10.02

These data agree with the majority of previous findings in that both men and women exhibit inaccuracies in self-reporting physical characteristics. In contrast, however, the majority of previous studies showed that both men and women tended to underreport weight and overreport height.

---

[1] Source: Jacobson, B. H., and DeBock, D. H. (2001). Comparison of body mass index by self-reported versus measured height and weight. *Perceptual and Motor Skills, 92,* 128–132. Copyright © 2001 by Perceptual and Motor Skills. Reprinted with permission.

Questions for Exercise 13

Part A: Factual Questions

1. What was the average age of the students?

2. Assuming that the distribution of age is normal, what percentage of students was between 17.0 and 22.8?

3. On the average, was the self-reported weight for men higher *or* lower than their measured weight?

4. For weight, did men *or* women have a larger average difference between self-reported weight and measured weight?

5. Did men *or* women have greater variability in their self-reported weight?

6. For men, was there greater dispersion for self-reported weight *or* for measured weight?

7. Assuming that the distribution of self-reported height for men is normal, the middle 68% of the men had scores between what two values?

8. Assuming that the distribution of self-reported weight for women is normal, what percentage of the women reported weights between 61.2 and 70.26?

9. Assuming that the distribution of self-reported weight for women is normal, what percentage of the women reported weights between 52.14 and 70.26?

10. Assuming that the distribution of self-reported height for women is normal, what percentage reported heights between 1.55 and 1.62?

11. Assuming that the distribution of measured weight for women is normal, the middle 68% of the women had scores between what two values?

Part B: Questions for Discussion

12. Before reading the results in the excerpt, would you have predicted that self-reports and measured values of height and weight would differ? Explain.

13. Two methods (self-report and a scale) were used to measure weight. In your opinion, is one method inherently better than the other? Explain.

14. The researchers note a difference in the pattern of results in this study and the pattern in the majority of other studies on this topic. Before reaching a conclusion on this topic, how interested would you be in examining the other studies? Explain.

Exercise 14   College Students' Procrastination

Mean, Standard Deviation, and Approximate 95% and 99.7% Rules

Statistical Guide

To review the mean, see the statistical guide for Exercise 9. To review the standard deviation, see the statistical guide for Exercise 12.

If you go out two standard deviation units of both sides of the mean, you capture approximately the middle 95% of the cases in a normal distribution. (The precise rule for capturing 95% of the cases is given in the next exercise.) If you go out three standard deviations on both sides of the mean, you capture 99.7% of the cases.

Excerpt from the Research Article[1]

The Procrastination Log [PL]…was administered both at intake [pretest to a one-hour individual counseling session on procrastination] and on outtake [posttest]. The PL is the most widely used instrument for assessing difficulties with procrastination…. It is a 9-item self-report questionnaire that measures procrastination-related behavior during the past week. The instrument lists nine common behaviors associated with procrastination (e.g., "I went out when I should have been studying") and asks participants to rate each item on a 7-point Likert-type scale ranging from 1 (*true*) to 7 (*false*). [Higher scores indicate a higher degree of procrastination.]

After participants agreed to be involved in the study, they were randomly assigned to one of three experimental groups…. [In the "same-attribution" group, the counselor agreed with the participants' attributions for their procrastination behavior such as bad luck, fate, family of origin, and biology. In the "no-attribution group," the counselor said that no attributions were necessary to find a solution to the problem. In the "different-attribution group," the counselor attributed the cause of procrastination to a cause other than the one the participant attributed it to.]

The major finding of the current study was that, contrary to prediction, [the group given the intervention that offered no causal explanation outperformed the other two groups].

Table 1   *Group means and standard deviations [of Procrastination Log (PL) scores]* *

	Intake		Outtake	
	M	*SD*	*M*	*SD*
Same-attribution group (*n* = 27)	43.5	8.6	37.4	9.1
No-attribution group (*n* = 28)	45.7	5.9	33.4	10.8
Different-attribution group (*n* = 27)	41.9	10.7	36.8	9.2

*Higher scores indicate greater self-reported procrastination.

---

[1] Source: Cook, P. F. (2000). Effects of counselors' etiology attributions on college students' procrastination. *Journal of Counseling Psychology, 47,* 352–361. Copyright © 2000 by the American Psychological Association. Reprinted with permission.

Questions for Exercise 14

Part A: Factual Questions

1. On the average, which group had higher scores on intake?

2. At intake, which group had the greatest variability in their scores?

3. At intake, which group had the least variability in their scores?

4. On the average, which group showed the greatest improvement from intake to outtake?

5. Assuming that the distribution of PL scores for the no-attribution group at intake is normal, between what two values did approximately the middle 95% of the participants lie?

6. Does the mean intake score for the same-attribution group fall within the range of the two scores that you gave as your answer to question 5?

7. Assuming that the distribution of PL scores for the no-attribution group at outtake is normal, between what two values did approximately the middle 95% of the participants lie?

8. Assuming that the distribution of scores for the no-attribution group at outtake is normal, between what two values did the middle 99.7% of the participants lie?

9. For the different-attribution group at outtake, what percentage of the students had scores between 9.2 and 64.4?

Part B: Questions for Discussion

10. If your work is correct, the interval you calculated for question 8 should be larger than the interval you calculated for question 7. Does this make sense? Explain.

11. If you are a procrastinator, do you think that a one-hour counseling session would be of help to you? Explain.

12. Some participants might report what they think the experimenters expect them to report. For example, after treatment for phobia of snakes, participants might report less phobia after treatment only because they believe that the experimenter would like to have that outcome. Do you think that this type of problem might be present in this particular study? Explain.

13. Is it important to know that the PL is the "most widely used instrument for assessing difficulties with procrastination"? Explain.

Exercise 15   Sense of Humor and Creativity

Mean, Standard Deviation, and Precise 95% and 99% Rules

Statistical Guide

To review the mean, see the statistical guide for Exercise 9. To review the standard deviation, see the statistical guide for Exercise 12.

If you go out 1.96 standard deviation units on both sides of the mean in a normal distribution, you capture 95% of the cases. (In the previous exercise, the approximate 95% rule was given.)

If you go out 2.58 standard deviation units on both sides of the mean in a normal distribution, you capture 99% of the cases.

Excerpt from the Research Article[1]

The subjects were 51 female and 35 male mental health professionals including social workers, psychologists, and psychiatrists. Their ages ranged from 24 to 57 years. These adults were predominantly Caucasians from urban, middle-class backgrounds.

All completed the humor and creativity measures in a single session. The Multidimensional Sense of Humor Scale…is a 29-item inventory on which subjects rate their agreement or disagreement with statements. [It taps recognition of humor, appreciation of humor, playfulness, and use of humor as an adaptive coping mechanism.]

On the Franck Drawing Completion Test [the measure of creativity], subjects are given a sheet of paper containing 12 incomplete line drawings that they are asked to complete any way they want. Drawings are scored for originality on a 6-point scale.

Results.—For the Multidimensional Sense of Humor Scale, the possible range of scores is from 0 to 96. The actual range for the [entire] present sample was 29 to 94, with a mean of 54.0 and a standard deviation of 6.3. The possible range of scores for the Franck Drawing Completion Test is from 0 to 60, with the actual range [for the entire sample] in this study of 12 to 56 ($M = 32.0$, $SD = 5.7$).

The subjects were then divided into low and high scoring groups on creativity…. The mean humor score for the high-scoring creativity group ($n = 71$) was 69.3 ($SD = 11.6$) and that for the low-scoring creativity group ($n = 15$) was 48.9 ($SD = 5.3$).

Humor and creativity seem associated in that both involve some risk taking and the production of unexpected and unusual responses.

Questions for Exercise 15

Part A: Factual Questions

1. Based on the fourth paragraph of the excerpt, which of the following value(s) tell(s) us about variability? (Circle one or more.) Explain your choice(s).

   A. 54.0   B. 6.3   C. 32.0   D. 5.7

---

[1] Source: Reproduced with permission of the authors and publisher from: Humke, C., & Schaefer, C. E. (1996). Sense of humor and creativity. *Perceptual and Motor Skills, 82,* 544-546. Copyright © Perceptual and Motor Skills 1996.

2. Assuming that the distribution of humor scores for the entire sample is normal, the middle 95% of the cases lies between what two values? (Use the multiplier given in the Statistical Guide for this exercise. Round your answer to one decimal place.)

3. Assuming that the distribution of humor scores for the entire sample is normal, the middle 99% of the cases lies between what two values? (Round your answer to one decimal place.)

4. Assuming that the distribution of scores on the Multidimensional Sense of Humor Scale for the entire sample is normal, what percentage of the cases lies between scores of 37.7 and 70.3?

5. Assuming that the distribution of scores on the Multidimensional Sense of Humor Scale for the entire sample is normal, what percentage of the cases lies between scores of 41.7 and 66.3? (Round your answer to one decimal place.)

6. Assuming that the distribution of humor scores for the *low-scoring creativity group* is normal, the middle 95% of the cases lie between what two scores? (Round your answer to one decimal place.)

7. Assuming that the distribution of humor scores for the *low-scoring creativity group* is normal, the middle 99% of the cases lie between what two scores? (Round your answer to one decimal place.)

8. In which group is there more variability in humor scores? Explain your answer.
   A. High-scoring creativity group.   B. Low-scoring creativity group.

9. Which group has the higher average humor score?
   A. High-scoring creativity group.   B. Low-scoring creativity group.

Part B: Questions for Discussion

10. The authors state that "humor and creativity seem associated." Does your answer to question 9 support this statement? Explain.

11. There were 71 subjects in the high-scoring creativity group and 15 subjects in the low-scoring creativity group. For purposes of comparing humor scores between these two groups, would it have been desirable to have equal numbers in both groups? Explain.

12. The authors of the excerpt did not state whether the distributions of scores were normally distributed. (This is true of most research writers in the social and behavioral sciences; it is traditional *not* to address normality in their research reports.) Does this pose a problem for consumers of research? Explain.

13. What is your opinion on measuring humor using a self-report inventory on which subjects rate their agreement or disagreement with statements? Speculate on other ways to measure humor.

Exercise 16 Problems of Students Absent from School

T Score

Statistical Guide

T scores have a mean of 50.00 and a standard deviation of 10.00 in the norm group on which a new test is standardized. Test makers prepare norms tables that allow us to convert a person's raw score (points earned) to the T score equivalents that would have been obtained if that person had been in the norm group. Likewise, we can compare the mean and standard deviation of another group to the mean of 50.00 and standard deviation of 10.00 in the norm group.

Since there are about three standard deviation units on both sides of the mean in a normal distribution, in practice, T scores can range from 20.00 to 80.00 (i.e., 3 x the standard deviation of $10.00 \pm$ the mean of 50.00).

Excerpt from the Research Article[1]

Subjects consisted of 44 adolescents (17 males and 27 females) 12 to 18 years of age [who had a] minimum of 20% absences from school in the 4 weeks prior to evaluation for the study. [On average, they missed 72% of full or partial school days.] Subjects were recruited through biannual mailings sent to middle, junior high, and high schools in the seven-county metropolitan area surrounding Minneapolis and St. Paul. Referrals were made by school personnel, physicians, mental health workers, and family members.

Child Behavior Checklist. The CBCL is a report measure that the parent completes about the child... There are eight scales of the CBCL (Withdrawn, Somatic Complaints [i.e., medical complaints about the body], Anxious/Depressed, Social Problems, Thought Problems, Attention Problems, Delinquent Behavior, and Aggressive Behavior). T scores of 70 (98th percentile) or greater are considered clinically significant.

In reporting their adolescents' symptoms on the CBCL, mothers endorsed the Somatic Complaints scale as having the highest mean ($T = 72.5 \pm 11.4$) (Table 4). The next highest group mean scores were on the Anxious/Depressed scale ($T = 70.4 \pm 10.7$) and the Withdrawn scale ($T = 69.8 \pm 10.6$).

Table 4 *CBCL scores reported by mothers*

Scales	T scores	
	Mean	SD
Withdrawn	69.8	10.6
Somatic Complaints	72.5	11.4
Anxious/Depressed	70.4	10.7
Social Problems	62.6	11.8
Thought Problems	60.0	8.7
Attention Problems	64.1	7.9
Delinquent Behavior	63.1	7.3
Aggressive Behavior	61.0	9.2

[1] Source: Bernstein, G. A., Massie, E. D., Thuras, P. D., Perwien, A. R., Borchardt, C. M., & Crosby, R. D. (1997). Somatic symptoms in anxious-depressed school refusers. *Journal of the American Academy of Child and Adolescent Psychiatry, 36,* 661–668. Copyright © 1997 by Williams & Wilkins. Reprinted by permission.

Questions for Exercise 16

Part A: Factual Questions

1. On which scale did the students in this study have the lowest average score?

2. On which scale did the students in this study have the most variability in their scores?

3. Suppose you were able to examine the scores of the individuals in this study. On which scale would you expect to find the largest differences among the scores of the students? Explain.

4. Is the mean score for the students in this study on the Anxious/Depressed scale above or below the mean for the norm group? Describe the basis for your answer.

5. On how many of the scales is the mean for the students in this study above the mean for the norm group?

6. On which scales is the mean score for students in this study in the "clinically significant" area? (See the excerpt for the definition of "clinically significant.")

7. Is it likely that some of the individual students in this study had "clinically significant" scores on the Withdrawn scale? Explain.

8. Is it likely that some of the individual students in this study had scores lower than 70 on the Somatic Complaints scale? Explain.

9. Assuming that the distribution of scores on the Delinquent Behavior scale for the students in this study is normal, the middle 68% had scores between what two values?

10. Assuming that the distribution of scores on the Delinquent Behavior score *for the students in the norm group* is normal, the middle 68% had scores between what two values?

Part B: Questions for Discussion

11. Compare your answers to questions 9 and 10. Based on the performance of the middle 68% on the Delinquent Behavior scale, would you be willing to say that the students in this study are clearly higher than those in the norm group? Explain.

12. In your opinion, would it be a good idea to arrange the scales in Table 4 alphabetically? Explain.

13. The authors repeated only some of the values from Table 4 in their discussion of the table. In your opinion, is this appropriate, or should they have discussed each of the values?

14. Many researchers report the means and standard deviations of raw scores (i.e., points earned) rather than derived scores such as *T* scores. Which is more helpful to you when interpreting the statistics? Explain.

Exercise 17   Cooperation and Friendship

Effect Size

Statistical Guide

In many experiments, we subtract the mean of the control group from the mean of the experimental group in order to determine the mean difference. However, we can get more information by dividing this difference by the standard deviation of the control group. The result (known as the effect size or *ES*) indicates the *number of standard deviations* by which the two groups differ; if the experimental group has a higher mean, it tells us by how many standard deviation units the experimental group exceeds the control group. Some researchers have suggested that an effect size of .5 standard deviation units or greater should be considered large (this is when the experimental group is one-half of a standard deviation above the control group). Further, they suggest that effect sizes of .3 to .5 are moderate, .1 to .3 are small, and less than .1 are trivial.[1]

Those of you who have studied *z*-scores may have noticed a parallel: A *z*-score tells us by how many standard deviation units an examinee differs from the mean of a group, while an effect size tells us by how many standard deviation units an experimental group mean differs from a control group mean.

Excerpt from the Research Article[2]

This article reports the results of a 2-year study of the cooperative elementary school model, which used cooperation as an overarching philosophy to change school and classroom organization and instructional processes. The components of the model include: using cooperative learning across a variety of content areas, full-scale mainstreaming of academically handicapped students, teachers using peer coaching, teachers planning cooperatively, and parent involvement in the school. The comparison schools continued using their regular teaching methods and curriculum.

The sample consisted of 1,021 students in second through sixth grades in five elementary schools of a suburban Maryland school district. Twenty-one classes in the two treatment schools were matched with 24 classes in the three comparison schools....

The social relations measure asked students to list the names of their friends in the class.... For learning-disabled students, the social relations measures were reanalyzed to determine the number of times each learning-disabled student was selected as a friend by his/her nonhandicapped peers [i.e., those peers who were not learning-disabled].

---

[1] These guidelines for interpreting effect sizes are drawn from Rosenthal, R., & Rosnow, R. L. (1984). *Essentials of behavioral research: Methods and data analysis*. New York: McGraw-Hill.

[2] Source: Stevens, R. J., & Slavin, R. E. (1995). The cooperative elementary school: Effects on students' achievement, attitudes, and social relations. *American Educational Research Journal, 32*, 321–351. Copyright © 1995 by the American Educational Research Association. Reprinted by permission of the publisher.

Table 3   *Classroom social relations, means, standard deviations, and effect sizes*

	CES [Cooperative]	Comparison [Control]	Effect size[1]
All students			
Number of friends			
Premeasure	5.91 (3.67)	5.89 (3.59)	+.01
Postmeaure	8.71 (3.82)	7.12 (3.62)	+.44
N (students)	411	462	
Gifted students			
Number of friends			
Premeasure	5.74 (2.23)	5.66 (2.54)	+.03
Postmeaure	8.92 (3.01)	7.59 (2.89)	+.46
N (students)	46	61	
Learning-disabled students			
Number of friends			
Premeasure	5.03 (2.81)	5.10 (3.04)	−.02
Postmeaure	9.11 (4.27)	6.29 (3.29)	+.86
Number of times picked by nonspecial education students [non-learning disabled]			
Premeasure	2.83 (1.76)	2.78 (1.54)	+.03
Postmeaure	5.83 (2.39)	3.81 (2.24)	+.90
N (students)	40	36	

[1]Effect size equals the difference in treatment means divided by the control standard deviation.

Questions for Exercise 17

Part A: Factual Questions

1. For gifted students on the postmeasure, what is the mean difference between the Cooperative group and the Comparison group? (Subtract the Comparison group mean from the Cooperative group mean.)

2. Divide your answer to question 1 by the standard deviation for the Comparison group on the postmeasure. What is your answer? Does it match any of the other values shown in the table? Explain.

3. For all students, how should you characterize the effect size on the premeasure? (See the Statistical Guide for this exercise before answering this question.)

   A. large   B. moderate   C. small   D. trivial

4. For all students, how should you characterize the effect size on the postmeasure? (See the Statistical Guide for this exercise before answering this question.)

   A. large   B. moderate   C. small   D. trivial

5. Compare your answers to questions 3 and 4. Does the comparison suggest that the Cooperative program is superior in promoting friendship among all students? Explain.

6. One of the effect sizes is negative. What does this tell us about the difference between the means of the two groups?

7. Consider the effect size on the postmeasure for "number of times picked by nonspecial education students." By how many standard deviation units does the Cooperative group exceed the Comparison group?

Part B: Questions for Discussion

8. Suppose the standard deviation for the Comparison group of gifted students had been 9.00 instead of 2.89 on the postmeasure. What would the effect size be? If your work is correct, it would be much smaller than the +.46 reported in the excerpt. Does it make sense that when there is a larger standard deviation (other things being equal), you get a smaller effect size? Explain.

9. In your opinion, which of the following is a better measure of number of friendships? Explain your choice.

   A. Number of friends named by special education students.
   B. Number of times special education students are picked by nonspecial education students.

10. To date, most researchers report means and standard deviations but do not report effect sizes, which is a relatively new statistic. In your opinion, would it be desirable for more researchers to report effect sizes? Explain.

Exercise 18 Measuring Functional Abilities of the Elderly

Scattergram

Statistical Guide

A scattergram (also known as a scatter plot or scatter diagram) depicts the relationship between two variables. For each case or participant, one dot is placed to show where the case stands on both variables. Patterns of dots from the lower left to the upper right indicate a direct relationship; patterns from the upper left to the lower right indicate an inverse relationship. The more scatter among the dots, the weaker the relationship.

Excerpt from the Research Article[1]

Performance-based measures of function [in elderly people], particularly activities of daily living…when collected in a standardized clinic setting, are presumed to be a meaningful reflection of the conduct of similar tasks as usually performed in the home. However, the extent to which performance on tasks conducted in a clinic setting under "idealized" conditions of lighting and clinic staff support reflects performance on the same task in the home…is not known.

The 97 participants were 67% female, 24% Black, and 60% aged 65–74…. A total of 20% of the participants in the sample had usual vision worse than 20/40. Between 95% and 100% of participants attempted each task in the home, except for the task of stair climbing, where 56 participants (58%) had no home staircase.

Tests:

Stair climb and descend. The number of seconds to climb up a set of stairs and the time to descend the same set of steps at a lighting level used routinely by the participant. In the clinic, participants are asked to climb seven steps set at a 32-degree incline. Data are presented as number of steps per second.

Plug insertion. The number of seconds to insert a plug into an electrical socket. A socket in the kitchen, preferably at waist level or higher, was selected. In the clinic, a board with a plug and socket is presented at eye level to a participant seated at a desk.

Telephone number look-up and dial. The participant was tested for the number of seconds to locate a telephone number on a standard page in a telephone book. A photocopied page of the local directory and the same number for all participants were used in the clinic setting. The number of seconds required to dial the number on a push-button phone…was recorded.

---

[1] Source: West, S. K., Rubin, G. S., Munoz, B., Abraham, D., & Fried, L. P. (1997). Assessing functional status: Correlation between performance on tasks conducted in a clinic setting and performance on the same task conducted at home. *Journal of Gerontology: Medical Sciences, 52A*, 4, M209–M217. Copyright © by the Gerontology Society of America.

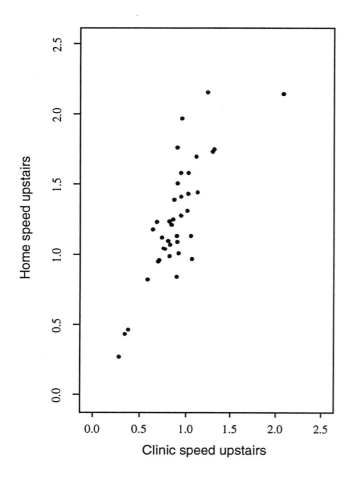

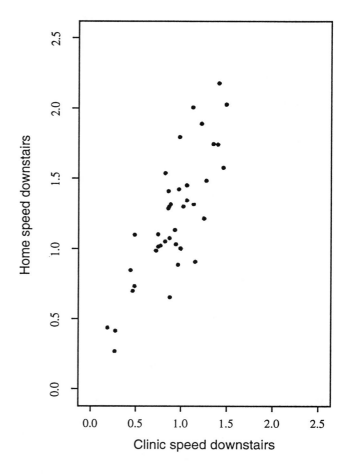

Figure 2   *Correlation between speed of climbing upstairs in the clinic versus in the home setting. (Data are in stairs per sec.)*

Figure 3   *Correlation between speed of climbing downstairs in the clinic versus in the home setting. (Data are in stairs per sec.)*

Questions for Exercise 18

Part A: Factual Questions

1. Are the four relationships depicted by the scattergrams direct or inverse? Explain.

2. Examine Figure 3 and find the participant with the best speed in the home. This person descended about how many stairs per second in the home?

3. Did the participant identified in question 2 have the best performance when tested in the clinic? Explain.

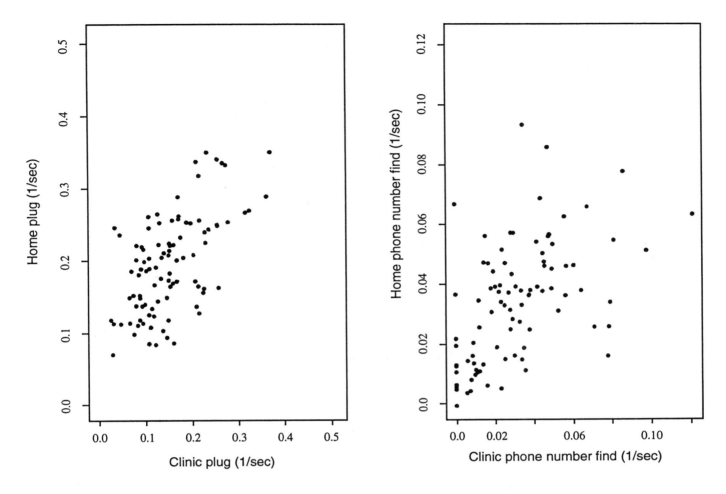

Figure 4 *Correlation between speed to insert a plug in a socket in the clinic versus in the home setting. (Data are 1/sec.)*

Figure 5 *Correlation between speed to find a telephone number in the clinic versus in the home setting. (Data are 1/sec.)*

4. Which one of the four scattergrams indicates the strongest relationship?

5. Which one of the four scattergrams indicates the weakest relationship?

6. Is Figure 3 or Figure 4 based on a larger number of participants? Explain.

7. The scores obtained in the clinic are listed on which axis in the scattergrams?
   A. The *x*-axis.   B. The *y*-axis.

8. In Figure 2, the person who had a clinic score of about 0.3 (the dot closest to the lower left-hand corner) had about what score when tested at home?

9. In Figure 2, the person who had a clinic score of about 2.2 (the dot closest to the upper right-hand corner) had about what score when tested at home?

Part B: Questions for Discussion

10. Compare your answers to questions 8 and 9. Do they suggest a direct relationship between the clinic scores and home scores for speed going upstairs? Explain.

11. Suppose you were assessing some elderly people for an important decision such as whether their functioning is adequate for them to continue to live at home without assistance. In your opinion, are the relationships in the figures strong enough to suggest that testing in a clinic alone would yield valid predictions of how well they function at home? (In other words, would it be sufficient to test them only in the clinic in order to make a decision about whether they can function adequately in the home?) Explain.

12. Many textbook authors suggest that the two axes for a scattergram should be about equal in length. The graphic artist who drew the scattergrams in the excerpt failed to follow this suggestion. That is, he or she made the *x*-axis shorter than the *y*-axis in each scattergram. In your opinion, is this failure important? Explain.

13. Are you surprised that the relationships in this excerpt are direct (i.e., positive)? Explain.

Exercise 19 Relationships Among Social Variables
Correlation Coefficient

Statistical Guide

A correlation coefficient indicates the strength and direction of a relationship between two variables. The most popular correlation coefficient is the Pearson r. When it is positive in value, the relationship is direct (that is, those with high scores on one variable tend to have high scores on the other variable *and* those with low scores on one variable tend to have low scores on the other). In a direct relationship, the closer r is to 1.00, the stronger the relationship; the closer it is to 0.00, the weaker the relationship.

When the value of the Pearson r is negative, the relationship is inverse (that is, those with high scores on one variable tend to have low scores on the other one). In an inverse relationship, the closer r is to -1.00, the stronger the relationship; the closer it is to 0.00, the weaker the relationship.

Background Notes

Aggression and withdrawal were measured by showing a picture of all classmates and asking each student to choose two classmates who best fit each descriptor. For *aggression*, a score was obtained for each child by summing the number of times he or she was selected for these descriptors: "gets into lots of fights," "loses temper easily," "too bossy," and "picks on other kids." For *withdrawal*, a score was obtained for each child by summing the number of times he or she was selected for these descriptors: "rather play alone than with others" and "very shy."

Social preference was assessed by asking each child to name three other children they would like most and like least for playing together, inviting others to a birthday party, and sitting next to each other on a bus. [Responses were scored in such a way that higher scores indicate greater social preference.]

Victimization by peers was measured by asking each child to nominate up to five other students who could be described as being made fun of, being called names, and getting hit and pushed by other kids. [Higher scores indicate greater victimization.]

Number of affiliative links was measured by asking, "You have probably noticed children in your class who often hang around together and others who are more often alone. Could you name the children who often hang around together?" [Higher scores indicate a larger number of affiliative links.]

Loneliness was measured with a 16-item questionnaire with higher scores indicating greater loneliness.

Perceived social acceptance and *behavior-conflict* were two aspects of self-concept measured with Harter's Self-Perception Profile for Children, which is a questionnaire. Higher scores reflect a better self-concept in each of the two domains.

Excerpt from the Research Article[1]

French Canadian children (393 girls, 400 boys; mean age = 115 months, range 8 to 10 years) participated in the study. The children attended third ($n = 315$), fourth ($n = 248$), and fifth ($n = 230$) grades in 10 elementary schools from a variety of socioeconomic backgrounds....

Table 1 presents the correlations among the measures considered in the present study.

Table 1 *Correlations among the social behavior, peer experiences, and self-perception measures*

Measure	1	2	3	4	5	6	7	8
1. Withdrawal	—							
2. Aggression	−.10	—						
3. Social preference	−.39	−.44	—					
4. Victimization by peers	.42	.53	−.68	—				
5. No. of affiliative links	−.35	.05	.35	−.21	—			
6. Loneliness	.29	.12	−.34	.34	−.18	—		
7. Perceived social acceptance	−.27	−.04	.28	−.26	.18	−.69	—	
8. Perceived behavior–conduct	.06	−.32	.17	−.17	−.06	−.35	.39	—

Questions for Exercise 19

Part A: Factual Questions

1. What is the value of the Pearson r for the relationship between withdrawal and aggression?

2. What is the value of the Pearson r for the relationship between loneliness and perceived social acceptance?

3. Is the relationship between withdrawal and loneliness direct or inverse? Explain the basis for your answer.

4. Is the relationship between loneliness and social preference direct or inverse? Explain the basis for your answer.

5. Which variable has the strongest relationship with withdrawal? Explain.

---

[1] Source: Boivin, M., & Hymel, S. (1997). Peer experiences and social self-perceptions: A sequential model. *Developmental Psychology*, *33*, 135–143. Copyright © 1997 by the American Psychological Association, Inc. Reprinted with permission.

6. Which variable has the weakest relationship with withdrawal? Explain.

7. The Pearson *r* for the relationship between withdrawal and loneliness indicates that those who tend to be more lonely tend to be

   A. more withdrawn.   B. less withdrawn.

8. Which of the following pairs of variables has the strongest relationship between them?

   A. Perceived social acceptance and loneliness
   B. Withdrawal and victimization by peers
   C. Number of affiliative links and aggression

9. Which of the following pairs of variables has the weakest relationship between them?

   A. Withdrawal and social preference
   B. Withdrawal and perceived social acceptance
   C. Withdrawal and perceived behavior–conduct

Part B: Questions for Discussion

10. In words (without using numbers), how would you describe the strength and direction of the relationship between withdrawal and aggression?

11. In words (without using numbers), how would you describe the strength and direction of the relationship between victimization by peers and social preference?

12. In your opinion, does it make sense that the Pearson *r* for the relationship between loneliness and perceived social acceptance is –.69? Explain.

13. In your opinion, does the Pearson *r* of –.27 for the relationship between withdrawal and perceived social acceptance prove that being withdrawn *causes* less perceived social acceptance? Explain.

Exercise 20   Academic Self-Efficacy and Social Goals
Correlation Coefficient and Coefficient of Determination: I

Statistical Guide

To review correlation coefficients, see the statistical guide for Exercise 19.

Consumers of research should compute coefficients of determination when interpreting correlation coefficients. The coefficient of determination is simply r squared; its symbol is r^2. For example, if $r = .50$, $r^2 = .50 \times .50 = .25$. If we multiply this by 100, we get the explained variance (also known as the amount of variance accounted for). In our example, the explained variance is .25 x 100 = 25%. This means that a Pearson r of .50 represents a relationship that is 25% higher than a Pearson r of 0.00 (that is, it is 25% better than no relationship). It is important to notice that a Pearson r of .50 is *not* 50% better than no relationship since the coefficient of determination must be used to arrive at the percentage.

Excerpt from the Research Article[1]

The participants in this study were 753 fifth-grade students (380 boys and 373 girls) from three school districts in southeastern Michigan. Students were recruited from 35 classrooms....

Perceived social efficacy. The measure of perceived social efficacy with peers includes items about students explaining their views, social group entry, and working out social problems with their classmates. The measure of self-efficacy perceptions of social interactions with the teacher was a parallel scale to the peers measure.

Intimacy goals. [This scale] includes items related to forming and maintaining positive intimate peer relationships, including close friendships and more general group acceptance.

Responsibility goals. [This scale] refers only to compliance with classroom norms and teacher requests.

Perceived academic self-efficacy. This scale refers to students' judgments of their capability to do the work in their current class and does not mention specific subject areas.

[For all measures, higher scores indicate more of the trait in question.]

Table 1   *Correlations among academic efficacy, social variables, gender, and GPA*

	1	2	3	4	5	6	7
1. Gender[a]	—						
2. Prior GPA[b]	.19	—					
3. Social efficacy: Peers	−.02	.20	—				
4. Social efficacy: Teacher	.09	.17	.26	—			
5. Intimacy goals	.22	.12	.10	.28	—		
6. Responsibility goals	.17	.09	.09	.50	.41	—	
7. Academic self-efficacy	−.04	.22	.26	.31	.17	.31	—

[a]Gender is coded boys = 0, girls = 1.
[b]GPA is coded E = 1, A+ = 13.

---

[1] Source: Patrick, H., Hicks, L., & Ryan, A. M. (1997). Relations of perceived social efficacy and social goal pursuit to self-efficacy for academic work. *Journal of Early Adolescence, 17*, 2, 109–128. Copyright © 1997 by Sage Publications, Inc. Reprinted by permission of Sage Publications.

Questions for Exercise 20

Part A: Factual Questions

1. What is the value of the Pearson *r* for the relationship between prior GPA and academic self-efficacy?

2. The Pearson *r* for the relationship between gender and prior GPA indicates that (Circle a choice and describe your reasoning.)

   A. girls tend to have a higher prior GPA.
   B. boys tend to have a higher prior GPA.

3. The Pearson *r* for the relationship between gender and intimacy goals indicates that (Circle a choice and describe your reasoning.)

   A. girls tend to have higher intimacy goals.
   B. boys tend to have higher intimacy goals.

4. Is there much of a relationship between gender and social efficacy with peers? Explain.

5. What is the value of the coefficient of determination for the relationship between academic self-efficacy and responsibility goals?

6. What percentage of the variance on academic self-efficacy is explained by the variance in responsibility goals?

7. What is the value of the coefficient of determination for the relationship between prior GPA and intimacy goals?

8. What percentage of the variance on intimacy goals is explained by the variance in prior GPA?

9. The strongest relationship in Table 1 is between what two variables?

10. What is the percentage of explained variance for the correlation coefficient that gave you the answer to question 9?

11. In words (without using numbers), how would you describe the strength and direction of the relationship between academic self-efficacy and gender?

12. The relationship between intimacy goals and responsibility goals is such that

    A. those who score higher on intimacy tend to score higher on responsibility.
    B. those who score higher on intimacy tend to score lower on responsibility.

Part B: Questions for Discussion

13. Would you have expected a stronger relationship between prior GPA and academic self-efficacy than reported in this study? Explain.

14. Gender is measured as a nominal variable (i.e., a categorical variable—not a continuous score variable). Examine your statistics textbook to see if the Pearson *r* is recommended for use with such variables. Write your findings here.

Exercise 21   New Parents Project

Correlation Coefficient and Coefficient of Determination: II

Statistical Guide

To review correlation coefficients, see Exercise 19. To review coefficients of determination and percentage of explained variance, see Exercise 20. In addition, you need to know that the interpretation of a coefficient of determination for a negative value of r is the same as for a positive value except that it applies to an *inverse* relationship. For example, .16 (a positive value) is the coefficient of determination for an r of −.40 (a negative value). Multiplying by 100, we learn that a value of −.40 is 16% away from 0.00 *in the negative direction.*

Coefficients of determination are always positive because of the squaring. Hence, we use positive values to interpret both positive and negative values of r.

Excerpt from the Research Article[1]

The 21 adolescent mothers were between the ages of 16 and 19 years.... [They had been recruited by Parents Project] personnel and then contacted by one of the researchers to solicit their participation. [The instruments that were used were]:

Revised UCLA Loneliness Scale.... Scores range from 20 to 80; the higher the score, the greater the loneliness.

Rosenberg Self-Esteem Scale. Self-esteem was measured using the 10-item...scale. Positively and negatively worded items were included in the scale to reduce the likelihood of response set. When five items are reverse-scored, higher scores indicate greater self-esteem.

Social Support Questionnaire—Short Form. Respondents list the people they can rely on for support in a given set of circumstances and indicate overall level of satisfaction with the support provided.

Center for Epidemiologic Studies Depression Scale for Children [CES-DC; with 20 items]. Scores higher than 15 indicate the presence of depressive symptomatology.

There was a negative relationship between depression and social support ($r = -.61$). Social support was positively associated with self-esteem ($r = .65$) and negatively associated with loneliness ($r = -.50$). Loneliness was correlated with depression ($r = .53$) and inversely correlated with self-esteem ($r = -.74$).

Questions for Exercise 21

Part A: Factual Questions

1. Which value of r in the excerpt represents the strongest relationship?

---

[1] Source: Hudson, D. B., Elek, S. M., & Campbell-Grossman, C. (2000). Depression, self-esteem, loneliness, and social support among adolescent mothers participating in the new parents project. *Adolescence, 335,* 445–453. Copyright © 2000 by Libra Publishers, Inc. Reprinted with permission.

2. Those who are high on depression tend to have what type of score on social support?

   A. A relatively high score.
   B. A relatively low score.

3. Those who are high on self-esteem tend to have what type of score on social support?

   A. A relatively high score.
   B. A relatively low score.

4. Which of the five correlation coefficients shown above has the smallest coefficient of determination? (Try to answer this question without performing any computations.)

5. What is the value of the coefficient of determination for the correlation you referred to in your answer to question 4?

6. What is the percentage of explained variance (variance accounted for) that corresponds to your answer to question 5?

7. To two decimal places, what is the value of the coefficient of determination for the relationship between loneliness and self-esteem?

8. To two decimal places, what is the percentage of explained variance (variance accounted for) that corresponds to your answer to question 7?

9. To two decimal places, for the relationship between depression and social support, what is the percentage of variance accounted for?

Part B: Questions for Discussion

10. Would you characterize any of the relationships in the excerpt as being "strong"? Explain.

11. Before this study was conducted, would you have hypothesized that the relationship between loneliness and depression would be direct *or* inverse? Explain.

12. Would you be willing to generalize the results of this study to all adolescent mothers in the country? Explain.

Exercise 22   Predicting College GPA
Multiple Correlation

Statistical Guide

To review correlation coefficients, see the statistical guide for Exercise 19.

A multiple correlation coefficient (R) indicates the extent to which a combination of variables predicts an outcome variable. (In the excerpt below, the researchers determine how well a combination of tests predict GPA. In this case, the multiple R tells us how well we can predict if we use the *combination* of tests—as opposed to using a single test.)

Squaring R yields the coefficient of determination, which is the proportion of variance in the outcome that is predicted by the combination of two or more predictor variables. This proportion is sometimes called the *explained variance* or *variance accounted for*.

Table 1 in this exercise shows how well one test (SCAT Total) predicts GPA. Table 2 shows how much R (and the corresponding value of R^2) are increased when additional tests are added into the formula for R.

Excerpt from the Research Article[1]

The D-48, a nonverbal test of general intelligence, is widely used in Europe and other areas of the world but is virtually unknown in the United States. In this study, the D-48 and the School and College Ability Tests (SCAT) [which are widely used in the U.S.] were administered to a sample of 250 community college students.... Of the 250 students, 145 were Mexican American and 105 were Anglo-American. [Means for ethnicity and gender were not significantly different.] A step-wise regression analysis indicated an R^2 of .36 with grade point average (GPA), with the D-48 entering as a second variable behind the SCAT Total. [Note that the SCAT consists of a verbal and a quantitative (numerical) part.]

Table 1   *Correlation coefficients between D-48, GPA, and SCAT scores*

	Samples			
	Mexican American		Anglo-American	
	Males	Females	Males	Females
D-48 and GPA	.41	.50	.45	.49
D-48 and SCAT Total	.57	.55	.56	.58
D-48 and SCAT Quantitative	.58	.52	.58	.55
D-48 and SCAT Verbal	.44	.47	.46	.49
SCAT Total and GPA	.54	.66	.52	.58
SCAT Quantitative and GPA	.56	.51	.53	.46
SCAT Verbal and GPA	.45	.58	.42	.51

[Note: The statistics in this table are Pearson *rs*—not multiple *Rs*.]

[1] Source: Domino, G., & Morales, A. (2000). Reliability and validity of the D-48 with Mexican American college students. *Hispanic Journal of Behavioral Sciences, 22*, 382–389. Copyright © 2000 by Sage Publications, Inc. Reprinted with permission.

Table 2   *Results of step-wise multiple regression*

Variable Entered	Multiple		Increase in R^2
	R	R^2	
1. SCAT Total	.537	.288	.288
2. D-48	.583	.340	.052
3. SCAT Numerical	.598	.358	.018
4. SCAT Verbal	.602	.362	.004

[Note: The statistics in this table are based on the total sample of Mexican American and Anglo-American students.]

Questions for Exercise 22

Part A: Factual Questions

1. In Table 1, which variable correlates most highly with GPA for Mexican American males?

2. In Table 1, which variable is the best predictor of GPA for Anglo-American males?

3. In Table 1, which variable is the least predictive of GPA for Mexican American females?

4. By how much does R^2 increase when the D-48 is added to the SCAT Total?

5. What is the value of the multiple correlation coefficient for when SCAT Total and D-48 are used together?

6. How much of the variance in GPA is accounted for by the combination of SCAT Total and D-48?

7. What is the value of the multiple correlation coefficient when all four variables are used in combination to predict GPA?

8. To two decimal places, what is the value of the coefficient of determination that corresponds to your answer to question 7?

Part B: Questions for Discussion

9. The best predictor of grades (i.e., GPA) for Mexican American females is SCAT Total ($r = .66$). How would you describe the strength of this relationship (e.g., "perfect," "extremely strong," and so on)?

10. In your opinion, is the increase provided by using the D-48 in conjunction with the SCAT Total large enough to justify having students take an additional test?

11. The SCAT is used three times in the multiple regression (as a total score, the verbal score, and the quantitative score). Does it make sense to use all three in such an analysis? Explain.

Exercise 23   Dieting, Age, and Body Fat
Linear Regression

Statistical Guide

To review scattergrams, see the statistical guide for Exercise 18.

In simple linear regression, a single straight line is mathematically fitted to describe the dots on a scattergram. The equation for any straight line is $y = a + bx$, where a is the intercept (i.e., the score value where the line meets the vertical axis) and b is the slope (i.e., the rate of change or the direction and angle of the line). Note that for a direct relationship, the slope will be positive in value; for an inverse relationship, it will be negative. After the best-fitting line has been mathematically determined for a particular scattergram, the values for x (the scores) can be inserted and the formula solved to obtain predicted values on y.

Background Notes

In the study described below, the subjects' body fat was pretested and posttested to assess the effects of a very low calorie diet. Hydrodensitometry measures body fat in water; bioelectrical impedance analysis (BIA) is an electrical measure of body fat.

Excerpt from the Research Article[1]

Seventeen subjects (nine women and eight men) from an outpatient, hospital-based treatment program for obesity volunteered.

Within 10 days after the baseline measures (i.e., pretest measures) were obtained, the subjects began the 12-week VLCD [very low calorie diet] portion of the...program.... At the end of the 12 weeks, all measurements were repeated in the laboratory.

Figure 1 demonstrates the relationship between age and percent of weight loss as fat. The correlation coefficient for this relationship was $r = -.49$.

Correlation between hydrostatic weighing and bioelectrical impedance was $r = .63$ for the 16 pretests and $r = .84$ for the 17 posttests. (Hydrostatic measurement of one male subject was not possible on the pretest because of discomfort in the water.) Correlation for the combined pretest and posttest trials ($n = 33$) was $r = .837$ (Figure 2).

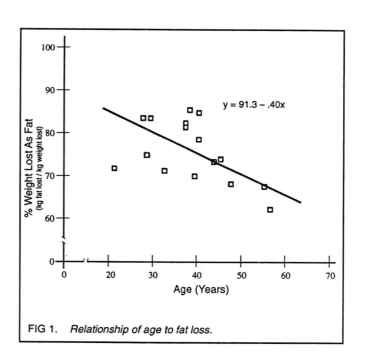

FIG 1.   *Relationship of age to fat loss.*

---

[1] Source: Burgess, N. S. (1991). Effect of a very low calorie diet on body composition and resting metabolic rate in obese men and women. *Journal of the American Dietetic Association, 91*, 430–434. Copyright © by The American Dietetic Association. Reprinted by permission.

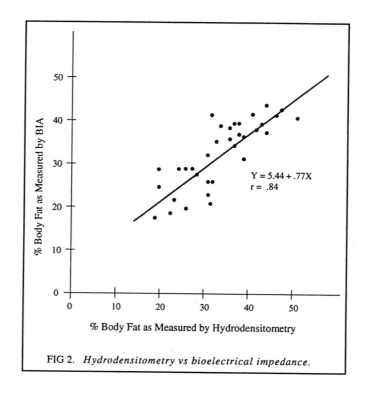

FIG 2. *Hydrodensitometry vs bioelectrical impedance.*

Questions for Exercise 23

Part A: Factual Questions

1. Which figure illustrates a direct relationship? Explain.

2. Which figure has a negative slope? Is the corresponding correlation coefficient also negative?

3. If the line in Figure 1 were extended to the left, at what value would it meet the vertical axis (i.e., *y*-axis)? (Give an answer that is precise to one decimal place.)

4. What is the value of the intercept in Figure 2?

5. Is the relationship depicted in Figure 1 or Figure 2 stronger? Explain.

6. Use the equation for the straight line shown in Figure 1 to predict the percentage of weight lost as fat for a person who is 60 years of age.

7. Use the equation for the straight line shown in Figure 1 to predict the percentage of weight lost as fat for a person who is 20 years of age.

8. Compare your answers to questions 6 and 7. Do your answers make sense considering that the relationship is inverse? Explain.

9. Use the equation for the straight line shown in Figure 2 to predict the percentage of body fat as measured by BIA for a person who has 50% body fat as measured by hydrodensitometry.

10. The equation in which figure will yield more accurate predictions? Explain.

Part B: Questions for Discussion

11. Suppose the slope in Figure 2 had been 1.50 instead of 0.77. With a slope of 1.50, would the line rise (from left to right) more steeply or less steeply than the line shown in Figure 2?

12. The correlation coefficient for the relationship between the two methods for measuring body fat is less than perfect. What does this tell us about the two methods of measurement?

13. Speculate on why we use a line based on the entire group to make predictions instead of using the values of individuals to make predictions. (For example, the person who is about 58 years of age in Figure 1 lost about 62% of weight as body fat. In the future, why don't we just predict that all those who are 58 years of age will lose 62% body fat?)

Exercise 24   Excerpts on Sampling

Bias in Sampling

Statistical Guide

In unbiased sampling, each member of a population has an equal chance of being included in a sample. Random sampling is the basic method used to obtain unbiased samples. Self-selection, volunteering, and any other nonrandom event (e.g., choosing people who happen to be convenient to serve as participants in a study) bias a sample. *Do not assume that a sample was drawn at random unless the researcher explicitly states this was done.*

Questions for Exercise 24

Directions: For each of the following excerpts from journal articles, indicate whether the sample is biased or unbiased and explain your choice. In some cases, you may answer "not sure" because there is insufficient information to make a judgment. If you answer "not sure" to an item, describe the additional information you would like to have before making a choice.

1. Researchers state, "To randomly select parents, teachers chose every third African American child on their class list and asked selected children to take the survey to their parents.... All participants were asked to mail their completed forms in self-addressed stamped envelopes to the researchers. We offered each participant $20 for their contribution to the study. Although most participants (80%) accepted payment, some participants (11%) returned the payment to the researchers.... Another 9% donated their payment to the school.... All completed surveys were individually returned by mail."[1]

2. Researchers state, "Nine counseling sites were invited to participate in the study. These sites were selected to provide a representative range of the types of locations at which counseling takes place. The locations were a university counseling center, a psychology training clinic, a women's center, a Lutheran Social Service center, two community mental health centers, and three private psychology practices. Two of the private practice sites declined to participate."[2]

3. Researchers state, "This study was conducted at a large, predominantly White southwestern university. On this campus, American Indians were the smallest racial and ethnic minority student group, consisting of only 2.3% of the student population.... Recruited through education and liberal arts classes, students who volunteered to participate in this study completed the research packet and returned it during the next class period. A total of 83 American Indian undergraduates returned completed survey packets. This sample represents about 10% of the undergraduate American Indian students on campus."[3]

4. Researchers report, "A systematic sample was drawn from the national membership lists of the American Society of Certified Accountants, National Association of Accountants, American Association of Women Accountants, and the Association of Government Accountants. An initial

name on each list was selected at random, and every *k*th name was thereafter selected. *K* was computed by dividing membership list length by the desired sample size and is defined as the sampling interval."[4]

5. A researcher states, "The participants were 38 early adolescent Latino boys and girls (20 boys, 18 girls). All students in the eighth grade (in this urban school district, most elementary schools included kindergarten through eighth grade) social studies classrooms in two midwestern urban elementary schools were invited to participate in the study. Students had been assigned randomly to the classrooms. Students who returned parental consent forms (approximately 60% of those invited) were interviewed individually [for this study]...."[5]

6. Researchers state that "Balancing the needs for efficiency and minimizing potential error, we developed a sampling plan to collect data from a representative stratified random sample of the 367 Wisconsin school districts with elementary schools...the stratification characteristic was socioeconomic status represented by the proportion of students eligible for the federal free and reduced-cost lunch program.... Districts were chosen within strata."[6]

7. A researcher states that "Data for this study came from a stratified random sample of persons aged 60 and over living in the community in Winnipeg, Manitoba. With assistance from the Manitoba Health Services Commission, the organization responsible for all health insurance claims (health insurance is universal in Manitoba), the sample was stratified by living arrangements. The final sample included 301 who were married and living with their spouse (included here are common-law marriages), 423 who live alone, and 560 who live with someone other than a spouse. Each of these samples is random and therefore representative for that group."[7]

8. A researcher states that "All program participants [in a self-paced precalculus course] were recruited for voluntary participation in a research study, which required filling out several psychological instruments mailed to the student pool. Of the total potential subjects, 280 (46.5%) returned the research packet. Of those 280, 6 did not have useable protocols and were dropped from this study, leaving a sample of 274."[8]

9. Researchers state that "Questionnaires were mailed to all female students ($n = 236$) and a random sample of male students ($n = 348$) [enrolled in a college of engineering] who met the following criteria: (a) entered college directly from high school, (b) declared a major in engineering upon enrolling, (c) were enrolled in the university as 3rd-, 4th-, or 5th-year students at the time of the study, and (d) were American citizens or residents. The final sample comprised all 278 eligible participants who returned their questionnaires through the mail."[9]

10. A researcher states, "Convenience samples of nutrition and nonnutrition majors were recruited from nutrition classes and various on-campus locations. The purpose of the study was explained, and the questionnaires were distributed, completed, and returned to the distribution sites."[10]

11. Researchers state, "Children were provided with consent forms in their classrooms to take to their parents for signatures. Teachers were asked to prompt the children at the end of the school day to remember their consent form and were asked to work for 100% consent rate. Stickers were provided on the return of the consent form (endorsed yes or no). Consent forms were obtained for 100 children (45.4%)."[11]

12. Researchers state that "To collect data with which to test the hypotheses, trained interviewers called 1,000 household telephone numbers selected randomly from a large city telephone directory. Three hundred and sixty-eight interviews were completed successfully with 'one of the adults' in the household. Others refused to participate, were not available by phone (in spite of at least one callback), and/or had their answering machines on. The gross response rate of 36.8 percent was determined to be acceptable for our exploratory study...."[12]

---

[1] Lambert, M. C., Puig, M., Lyubansky, M., Rowan, G. T., & Winfrey, T. (2001). Adult perspectives on behavior and emotional problems in African American children. *Journal of Black Psychology, 27,* 64–85.

[2] Rose, E. M., Westefeld, J. S., & Ansley, T. N. (2001). Spiritual issues in counseling: Clients' beliefs and preferences. *Journal of Counseling Psychology, 48,* 61–71.

[3] Gloria, A. M., & Kurpius, S. E. R. (2001). Influences of self-beliefs, social support, and comfort in the university environment on the academic nonpersistence decisions of American Indian undergraduates. *Cultural Diversity and Ethnic Minority Psychology, 7,* 88–102.

[4] Day, V. D., & Bedeian, A. G. (1991). Work climate and Type A status as predictors of job satisfaction: A test of the interactional perspective. *Journal of Vocational Behavior, 38,* 39–50.

[5] Yowell, C. M. (2000). Possible selves and future orientation: Exploring hopes and fears of Latino boys and girls. *Journal of Early Adolescence, 20, 245–280.*

[6] Graue, M. E., & DiPerna, J. (2000). Redshirting and early retention: Who gets the "gift of time" and what are its outcomes? *American Educational Research Journal, 37,* 509–534.

[7] Chappell, N. L. (1991). Living arrangements and sources of caregiving. *Journal of Gerontology: Social Sciences, 46,* 1–8.

[8] Parker, W. D. (1996). Psychological adjustment in mathematically gifted students. *Gifted Child Quarterly, 40,* 154–157.

[9] Schaefers, K. G., Epperson, D. L., & Nauta, M. (1997). Women's career development: Can theoretically derived variables predict persistence in engineering majors? *Journal of Counseling Psychology, 44,* 173–183.

[10] McArthur, L. H. (1995). Nutrition and nonnutrition majors have more favorable attitudes toward overweight people than personal overweight. *Journal of the American Dietetic Association, 95,* 593–596.

[11] Franz, D. Z., & Gross, A. M. (2001). Child sociometric status and parent behaviors: An observational study. *Behavior Modification, 25,* 3–20.

[12] Donthu, N., & Gilliand, D. (1996). Observations: The infomercial shopper. *Journal of Advertising Research, 36,* 69–76.

Exercise 25   Treatment of Depression and Anxiety

Standard Error of the Mean and 95% Confidence Interval

Statistical Guide

The standard error of the mean (SE_M) is a margin of error to allow for when estimating the population mean from a sample drawn at random from a population. It is an allowance for chance errors created by the random sampling. When a 95% confidence interval (CI) for a mean is reported, we can have 95% confidence that the *true mean* (i.e., the mean we would get if we could eliminate sampling errors) lies within the interval. For example, if we tested a group of children with a mathematics test and got a mean of 55.00 and a 95% CI of 52.00–57.00, we could have 95% confidence that the true mean lies between these two values.

When we report a single value such as a mean based on a sample, the value is sometimes called a *point estimate* (i.e., a single point estimated to be the mean for the population based on a sample). When we report a confidence interval, we are said to be reporting an *interval estimate* (i.e., a range of values that estimate the mean for a population).

Excerpt from the Research Article[1]

Twenty-nine adult psychiatric outpatients constituted the sample. They completed a 12-session group cognitive therapy program [in small groups]. Group sizes ranged from 3 to 6 members.

The instruments used in this program included the Beck Depression Inventory (BDI)..., [which] is a widely used instrument in the evaluation of depression.... Anxiety was measured with the Beck Anxiety Inventory (BAI), a 21-item self-report instrument that measures anxiety severity for the past week, including the day of completion. The Dysfunctional Attitudes Scale (DAS)...measures identification with attitudes associated with depressive disorders....

From the pre- to posttest, the average change on the BDI was a decrease of 8.42 points.... BAI scores fell an average of 4.65 points.... For the DAS, the average change from pre- to posttest was 21.42 points. [The 95% confidence intervals for these mean change scores are given in Table 1 below.]

Table 1 *Means and standard deviations for outcome measures*

Measure	Pretest		Posttest		95% CI
	M	*SD*	*M*	*SD*	
BDI	21.69	10.12	13.27	9.27	4.59–12.25
BAI	17.23	11.55	12.58	9.56	0.17–9.14
DAS	156.81	33.72	135.38	35.93	6.46–36.39

Note. N = 26. CI = Confidence interval; BDI = Beck Depression Inventory; BAI = Beck Anxiety Inventory; DAS = Dysfunctional Attitudes Scale.

---

[1] Source: Kush, F. R. & Fleming, L. M. (2000). An innovative approach to short-term group cognitive therapy in the combined treatment of anxiety and depression. *Group Dynamics: Theory, Research, and Practice, 4,* 176–183. Copyright © 2000 by the Educational Publishing Foundation. Reprinted by permission.

Questions for Exercise 25

Part A: Factual Questions

1. The anxiety scores fell an average of how many points?

2. For your answer to question 1 (the *point estimate*), what is the 95% confidence interval (the *interval estimate*)?

3. If you were reporting a point estimate for the mean change on the DAS, what value or values should you report?

4. If you were reporting an interval estimate for the mean change on the BDI, what value or values should you report?

5. For which measure does the interval estimate come closest to indicating zero change? Explain the basis for your answer.

6. Which confidence interval is largest?

7. Which confidence interval is smallest?

8. In light of the 95% confidence intervals, does the treatment appear to be effective? Explain.

Part B: Questions for Discussion

9. In your opinion, do the point estimates given in the text or the interval estimates given in the table provide better information about the effectiveness of the treatments? Explain.

10. In your opinion, is the sample size adequate for a study of this type? Explain.

11. Suppose a fourth variable had been measured, and the 95% confidence interval for it was −2.11 to 8.76. What would the negative tell you about the effectiveness of the treatment?

Exercise 26   Social Traits of Hearing-Impaired Students

Standard Error of the Mean and 68%, 95%, and 99% Confidence Intervals

Statistical Guide

To review the standard error of the mean, see the statistical guide for Exercise 25. Most authors of research reports in journal articles do not report the standard error of the mean. However, they usually give you enough information so that you can compute it using the formula shown in question 1 below.

For a 68% confidence interval, simply add the standard error of the mean to the mean and then subtract it from the mean. For a 95% confidence interval, first multiply the standard error of the mean by 1.96 and then add and subtract. For a 99% confidence interval, multiply by 2.58 before adding and subtracting. The intervals obtained using these multipliers are precise only with large samples; they become rather precise with 60 or more cases.

Excerpt from the Research Article[1]

Participants were 220 hearing-impaired adolescents enrolled in 15 large public school programs…throughout the United States and in one Canadian metropolitan area. Participants were in the high school part of the program.

Participation. Participation was tapped with four subscales: (a) One six-item subscale included items referring to participation in the classroom and in school with hearing-impaired peers (e.g., class related: "In my mainstream classes, I talk with hearing-impaired students"; school related: "I have lunch with hearing-impaired friends"). Students responded on a 5-point scale.

A second six-item subscale was identical to the first, except that the word *hearing* was substituted for *hearing-impaired* (e.g., "In my mainstream classes, I talk with hearing students").

An additional seven-item subscale dealt with participation in out-of-school social activities with hearing-impaired friends (e.g., "Go to parties at hearing-impaired friends' homes")…. A second seven-item subscale was identical to the preceding one, except *hearing* was substituted for *hearing-impaired* (e.g., "Go to parties at hearing friends' homes").

Emotional Security. Emotional security was tapped with two subscales: (a) A seven-item subscale included items referring to hearing-impaired students (e.g., "When I'm with hearing-impaired students my age, I feel nervous")…. (b) A second seven-item subscale was identical to the preceding one, except for the substitution of *hearing* for *hearing-impaired* (e.g., "When I'm with hearing students my age, I feel nervous")….

---

[1] Source: Stinson, M. S., Whitmire, K., & Kluwin, T. N. (1996). Self-perceptions of social relationships in hearing-impaired adolescents. *Journal of Educational Psychology, 88.* The excerpt appears on pages 132–143. Copyright © 1996 by the American Psychological Association, Inc. Reprinted by permission.

Table 1 *Means and standard deviations*

Subscale	M	SD
School Participation		
With hearing-impaired students	19.5	5.1
With hearing students	17.4	5.1
Social Activities		
With hearing-impaired students	18.5	5.6
With hearing students	19.0	5.5
Emotional Security		
With hearing-impaired students	22.2	3.7
With hearing students	21.6	3.8

Note: *N* for all subscales was 220. [Higher averages indicate more of the trait in question.]

Questions for Exercise 26

Part A: Factual Questions

1. What is the standard error of the mean for school participation with hearing-impaired students? Calculate it using the following formula. Round your answer to two decimal places.

$$SE_M = \frac{SD}{\sqrt{N}}$$

2. What are the limits of the 68% confidence interval for school participation with hearing-impaired students? Calculate your answer to two decimal places and round it to one decimal place.

3. What are the limits of the 95% confidence interval for school participation with hearing-impaired students? Calculate your answer to two decimal places and round it to one decimal place.

4. What are the limits of the 95% confidence interval for school participation with hearing students? Calculate your answer to two decimal places and round it to one decimal place.

5. Compare your answers to questions 3 and 4. Do the two 95% confidence intervals overlap (i.e., are any of the values in one of the intervals included in the other)? (Note that when two intervals do *not* overlap, we can be confident that there is a reliable difference between the means.)

6. We can have 68% confidence that the true mean for social activities with hearing-impaired students lies between which two values? Calculate your answer to two decimal places and round it to one decimal place.

7. We can have 95% confidence that the true mean for social activities with hearing-impaired students lies between which two values? Calculate your answer to two decimal places and round it to one decimal place.

8. Compare your answers to questions 6 and 7. If your work is right, the interval for question 7 is larger than the interval for question 6. Does this make sense? Explain.

9. Calculate the limits of the 99% confidence interval of the mean for social activities of hearing-impaired students. (Note that you calculated the 68% and 95% confidence limits for this variable in questions 6 and 7 above.)

Part B: Questions for Discussion

10. Keep in mind that each of the means and standard deviations in Table 1 is based on an *N* of 220. Without performing any calculations, can you determine which variable in Table 1 has the smallest standard error of the mean associated with it? Explain.

11. Compare the two means for each of the three main variables in Table 1. Are all the differences in the direction you would have predicted before reading the excerpt? Explain.

12. Briefly explain what caused the errors being accounted for by the confidence intervals for the means.

Exercise 27　Sexual Harassment in the Navy

t Test for Independent Groups: I

Statistical Guide

The *t* test is often used to test the significance of the difference between two means. It yields a value of *p*, which indicates the probability that chance or random sampling errors created the difference between the means. Most researchers declare a difference to be significant when *p* is equal to or less than .05. The lower the probability, the more significant the difference. Thus, a *p* of .01 is more significant than a *p* of .05. When a researcher declares a difference to be statistically significant, he or she is rejecting the null hypothesis. Note that *ns* means not significant.

Excerpt from the Research Article[1]

Women officers and women enlisted personnel [in the Navy] were categorized into two groups based on their survey responses: (1) sexually harassed and (2) not sexually harassed. Sexual harassment was reported by 182 women officers and not reported by 385 women officers. Of the women enlisted, 436 were sexually harassed and 582 were not. *T* tests were conducted between women who were harassed and those who were not harassed… [See Table 4.]

Table 4　*Mean responses to individual career experiences/Navy satisfaction items[a]*

Question	Those harassed	Those not harassed	*t*
1. I would recommend the Navy to others.			
Officers	3.66	3.98	3.35^b
Enlisted	3.31	3.60	3.76^b
2. I am satisfied with my rating.			
Officers	3.87	3.96	*ns*
Enlisted	3.24	3.56	4.02^b
3. I plan to leave the Navy because I am dissatisfied.			
Officers	3.58	3.87	2.78^c
Enlisted	3.17	3.67	5.89^b
4. My experiences have encouraged me to stay in the Navy.			
Officers	2.64	3.09	4.19^b
Enlisted	2.24	2.58	4.56^b
5. This command provides the information people need to make decisions about staying in the Navy.			
Officers	2.95	3.34	3.57^b
Enlisted	2.71	3.00	3.80^b
6. In general, I am satisfied with the Navy.			
Officers	3.48	3.88	4.44^b
Enlisted	3.29	3.68	5.41^b
7. I intend to stay in the Navy for at least 20 years.			
Officers	3.32	3.64	2.67^c
Enlisted	2.66	3.22	5.63^b

[a]Scores could range from 1 to 5. Higher scores indicate more positive perceptions. Question 3 is reverse scored; for this question, higher scores also indicate more positive perceptions.

[b]$p \le .001$

[c]$p \le .01$

[1] Source: Newell, C. E., Rosenfeld, P., & Culbertson, A. L. (1995). Sexual harassment experiences and equal opportunity perceptions of Navy women. *Sex Roles, 32*, 159–168. Reprinted by permission from Plenum Publishing.

The significant differences found on questions pertaining to satisfaction with the Navy and intent to serve for at least 20 years suggest that…sexual harassment may negatively impact long-range outcomes in areas such as retention and turnover.

Questions for Exercise 27

Part A: Factual Questions

1. Which group of officers had a higher average score in response to item 1 in Table 4?

2. Is the difference between the two groups of officers on item 1 in Table 4 statistically significant?

3. Using conventional standards, should we reject the null hypothesis for the officers on item 1 in Table 4?

4. The difference is *not* statistically significant for which pair of means in Table 4?

5. Should the null hypothesis be rejected for the officers on item 2 in Table 4?

6. How many of the differences in Table 4 are statistically significant at the .01 level but not significant at the .001 level?

7. How many of the differences in Table 4 led to rejection of null hypotheses at the .001 level?

8. Both differences on item 7 in Table 4 are statistically significant. Which one is significant at a higher level?
   A. The difference for officers   B. The difference for enlisted personnel

9. Consider the results for item 3 in Table 4. For which group can we reject the null hypothesis with greater confidence (i.e., for which group is there a smaller chance of making a Type I error)?
   A. Officers   B. Enlisted personnel

10. The first difference at the top of Table 4 is statistically significant at the .001 level. Is it also significant at the .05 level? (Note: If you have a statistics textbook, examine the table of critical values of *t* to help you determine the answer to this question.)

11. The first difference at the top of Table 4 is statistically significant at the .001 level. Is it also statistically significant at the .01 level?

12. Consider the three statistics in the row at the bottom of Table 4 (i.e., 2.66, 3.22, and 5.63). Which one is an inferential statistic? To what group (or branch) of statistics do the other two belong?

Part B: Questions for Discussion

13. Consider the first *t* test at the top of Table 4. For consumers of research, which of the following is more useful: (a) the value of *t*, which is 3.35, or (b) the value of *p*, which is $p \le .001$? (In other words, if you were reading the excerpt for the first time, which value would give you more useful information?) Explain.

14. In your opinion, would it have been a good idea for the researchers to also report the standard deviations associated with each mean in Table 4? Explain.

15. The subtitle of this exercise is "*t* Test for Independent Groups: I." If you have a statistics textbook, look up this term and describe when it is appropriate to use an "*independent t test*." (Note: In some books, it is called a *t* test for uncorrelated data.)

Exercise 28   Gender Differences in Computing

t Test for Independent Groups: II

Statistical Guide

To review the *t* test, see the statistical guide for Exercise 27.

Excerpt from the Research Article[1]

The respondents consisted of 150 students…. The student groups were formed by the college administrators at the beginning of the fall semester, and each group consisted of six students. The students remained in the same group for one year, and they worked together in several courses. In the computer course, the students worked jointly on a group project during the spring semester.

Self-Efficacy [in Computing]. The students were asked to rate how confident they were in performing each of four [computer] tasks on a 5-point scale ranging from "no confidence at all" to "complete confidence."

Previous Computer Experience. The students were asked to indicate on a 5-point scale to what extent they had worked with word processing, spreadsheet programs, programming, or computer games before attending college.

Previous Encouragement to Work with Computers. The subjects rated on a 5-point scale ranging from 1 to 5 the extent to which their decision to use computers had been influenced by parents, school teachers, and friends.

Group Processes. The students were asked to assess the activity level in the group, and to indicate to what extent they gave and received task-related help. These three questions were answered by indication on a 5-point scale ranging from "very small extent" to "very great extent."

Cooperation. The students were asked…what percentage of the time they spent in front of the computer consisted of cooperation with one or several fellow students….

Table 1   *t tests: Gender differences*

Variables	Female Students ($n = 63$)		Male Students ($n = 87$)		*t*	*p*
	Mean	SD	Mean	SD		
Self-efficacy in computing	10.7	2.8	12.8	3.4	4.32	.000*
Previous computer experience	8.5	2.8	10.4	3.5	3.79	.000*
Previous computer encouragement	5.4	2.1	6.7	2.7	3.42	.001*
Giving task-related help	2.6	.9	2.9	1.1	2.46	.015*
Receiving task-related help	3.2	.8	2.9	.9	−2.00	.048*
Level of activity in the groups	3.1	.7	3.3	.8	1.36	.180
Cooperation working with computers	61.0	29.9	44.1	29.4	−3.34	.001*

*$p < .05$

---

[1] Source: Busch, T. (1996). Gender, group composition, cooperation, and self-efficacy in computer studies. *Journal of Educational Computing Research. 15*, 125–135. Reprinted with permission from Baywood Publications, Amityville, NY.

Questions for Exercise 28

Part A: Factual Questions

1. The females have a higher average than the males on which two variables?

2. Is the difference on giving task-related help statistically significant at the .05 level?

3. Based on your answer to question 2, should the null hypothesis for giving task-related help be rejected at the .05 level?

4. Is the difference on giving task-related help statistically significant at the .01 level?

5. What is the precise probability that random sampling errors produced the difference between the means of female and male students on previous computer experience?

6. The value of p for cooperation working with computers is .001. What does this mean?
   A. 1 out of 10   B. 1 out of 100   C. 1 out of 1,000   D. 1 out of 10,000

7. Six of the seven values of p in the table have asterisks that refer the reader to a footnote. Explain why one of the values does not have an asterisk.

8. Which of the following differences is significant at a higher level (i.e., is more statistically significant)? Explain the basis for your answer.

   A. The difference on previous computer encouragement
   B. The difference on giving task-related help

9. Which of the following differences is significant at a lower level? Explain the basis for your answer.

   A. The difference on receiving task-related help
   B. The difference on cooperation working with computers

10. Why are two of the values of *t* negative? (Hint: If you have a statistics textbook, examine the formula for *t* to determine why it might produce a negative value.)

Part B: Questions for Discussion

11. The researcher reported precise values of *p* for each of the *t* tests (for example, a value of .048 for receiving task-related help). To what value of *p* did the researcher compare each of these to determine statistical significance?

12. If you have a statistics textbook, examine the discussion of the *t* test and the related table of critical values. Would it be possible for you to get a precise probability level such as .048 using only your textbook? Explain.

13. If you had been asked to make predictions about the outcomes of this study prior to the data collection, would you have predicted the types of gender differences reported in Table 1? Explain.

Exercise 29   Empathy Among Undergraduates

t Test for Independent Groups: III

Statistical Guide

To review the *t* test, see the statistical guideline for Exercise 27. Note that if a value of *t* in a table is not explicitly marked as being statistically significant, it is presumed to be insignificant.

Excerpt from the Research Article[1]

Undergraduate students (337 men and 353 women) were recruited for this research from psychology classes. Their ages ranged from 18 to 20 years.

Empathy was measured with the Interpersonal Reactivity Index.... The Index consists of 28 items covering four empathy subscales, i.e., Perspective-taking, e.g., "I sometimes try to understand my friends better by imagining how things look from their perspective," Fantasy, e.g., "I really get involved with the feelings of the characters in a novel," Empathic Concern, e.g., "I often have concerned feelings for people less fortunate than me," and Personal Distress, e.g., "Being in a tense emotional situation scares me." Responses were anchored by strongly disagree (1) and strongly agree (4).

Table 1 *Means and standard deviations of empathy subscales by sex*

| Empathy subscale | Men | | Women | | df | t |
	M	SD	M	SD		
Perspective-taking	17.89	3.64	18.37	3.52	685	1.75
Fantasy	20.80	4.13	21.89	3.95	684	3.53**
Empathic Concern	20.27	3.41	20.78	3.04	661	2.10*
Personal Distress	17.64	3.70	19.15	3.67	683	5.36***
Total scale	76.60	9.14	80.19	8.88	673	5.18***

* $p < .05$. ** $p < .01$. *** $p < .001$.

Questions for Exercise 29

Part A: Factual Questions

1. The first *t* test (in the first row of values of Table 1 with $t = 1.75$) compared which two values?

   A. 17.89 and 3.64   B. 17.89 and 18.37   C. 3.64 and 3.52

2. Was the null hypothesis for the first value of *t* rejected? Explain.

---

[1] Source: Ishikawa, T., & Uchiyama, I. (2000). Relations of empathy and social responsibility to guilt feelings among undergraduate students. *Perceptual and Motor Skills, 91,* 1127–1133. Copyright © 2000 by Perceptual and Motor Skills. Reprinted with permission.

3. The second value of *t* (i.e., 3.53) refers to the comparison of what?

   A. Two means.   B. Two standard deviations.

4. What is the probability that the third null hypothesis is true (i.e., for the null hypothesis that says men and women, on the average, are not truly different in empathic concern)?

5. Using conventional standards, should the null hypothesis for personal distress be rejected? Explain.

6. Using conventional standards, should the difference for personal distress be declared to be statistically significant? Explain.

7. Using conventional standards, should the difference for the total scale be declared to be statistically significant? Explain.

8. The researchers mention the .05, .01, and .001 levels in the excerpt. Which represents the highest level of significance?

9. The difference for which of the following is significant at a higher level?

   A. Fantasy.   B. Emphatic concern.

Part B: Questions for Discussion

10. If you had planned this study, would you have expected to find (i.e., hypothesized to find) significant differences between women and men? Explain.

11. Consider the statistics for the total scale in Table 1. For this scale, means and standard deviations are given. Also, we find that the null hypothesis was rejected at the .001 level. In light of this evidence, does it seem likely that all the men in this study are lower than all the women in terms of their empathy scores?

12. The values of *df* in the table are not all the same. If you have a statistics textbook, look up how *df* is computed. What does the information you located tell you about the data used to compute the statistics in the table?

13. The excerpt gives four sample items from the empathy measure. Do they give you a good idea of what the researchers mean by "empathy"?

Exercise 30   Therapy for Users of Illicit Drugs

t Test for Dependent Groups

Statistical Guide

The *t* test for dependent groups has the same purpose as the one for independent groups (see the statistical guide for Exercise 27). When each member of one group is paired or matched with a member of the other group, we have dependent data and should use the *t* test for dependent groups, which is also known as the *t* test for correlated data or the *t* test for paired data. Note that in the excerpt, the abbreviation *Ss* stands for *subjects*.

Excerpt from the Research Article[1]

The final study sample consisted of 74 *Ss*.... Approximately three-quarters of *Ss* were male, about three-quarters were adult, one-half were not employed or attending school, about one-third were mandated to obtain counseling by a public agency, and about one-eighth were minority persons.... Over one-half used cocaine, and about three-quarters used marijuana. [In addition, 9.5% used benzodiazipine, 9.5% used hallucinogens, 4.1% used barbiturates, 4.1% used opiates, and small percentages used other drugs.]

Subjects were assigned randomly to either a Supportive-discussion counseling program or to a directive Behavioral program after a 1-month pre-treatment assessment period. The three principal components of the Behavioral program were: (1) stimulus control, including competing response training; (2) an urge control procedure for interrupting incipient drug use urges, thoughts, or actions; and (3) behavioral contracting, especially between youth and their parents.

Figure 1 shows drug usage in terms of the mean number of days of drug use during the 1-month pre-treatment, the last month of treatment, and the follow-up month. For the Supportive Counseling *Ss*, a within groups paired *t* tests comparison showed that drug usage was unchanged from pre-treatment to end of treatment ($t = 0.23$, $p = 0.82$), and increased slightly, but non-significantly ($t = 0.53$, $p = 0.60$) from the time of treatment termination to the follow-up period, remaining statistically unchanged at follow-up from pre-treatment ($t = 0.19$, $p = 0.44$). For the Behavioral Counseling *Ss*, drug usage decreased substantially from pre-treatment to the end of treatment ($t = 4.28$, $p < 0.001$), with a slight, nonsignificant ($t = 0.92$, $p = 0.72$) further decrease from the time of treatment termination to the follow-up period, the decrease from pre-treatment to follow-up remaining statistically significant ($t = 4.42$, $p < 0.001$). [See Figure 1 on the next page.]

Questions for Exercise 30

Part A: Factual Questions

1. As is customary in journal articles, the researchers did not state null hypotheses. Write a statement of the null hypothesis for the first significance test in the excerpt.

---

[1] Source: Reprinted from: Azrin, N. H., Acierno, R., Kogan, E. S., Donohue, B., Besalel, V. A., & McMahon, P. T. (1996). Follow-up results of supportive versus behavioral therapy for illicit drug use. *Behaviour Research and Therapy*, *34*, 41–46 with kind permission from Elsevier Science Ltd., The Boulevard, Langford Lane, Kidlington 0X5 1GB, UK. Copyright © 1996 by Elsevier Science Ltd.

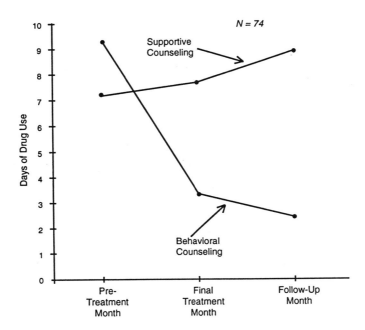

Figure 1 *Mean number of days of drug use during the one-month period previous to treatment, the final month of treatment, and the follow-up month for the Supportive Counseling and BehavioralCounseling programs. Treatment was for a mean of 8 months; follow-up occurred a mean of 9 months after treatment termination.*

2. Should the null hypothesis that you wrote from question 1 be rejected? Explain.

3. Would it be appropriate to reject the null hypothesis associated with the second *t* test in the excerpt? Explain.

4. In the excerpt, the researchers report that $p = .72$ for the decrease from treatment termination to the end of treatment for the Behavioral Counseling subjects. What does the value of p indicate?

   A. Odds are 72 out of 100 that biased procedures account for the difference.
   B. Odds are 72 out of 100 that the behavioral treatment was effective.
   C. Odds are 72 out of 100 that the behavioral treatment was not effective.
   D. Odds are 72 out of 100 that sampling error accounts for the difference.

5. The last difference discussed in the excerpt was statistically significant at the .001 level. Was it also significant at the .05 level? (Hint: You should be able to answer this question based on your knowledge of the relationships among significance levels. If not, consult a table of critical values of *t* in a statistics textbook.)

6. Four of the values of *t* in the excerpt are less than 1.00. If you have a statistics textbook, consult the table of critical values of *t* to determine whether a value of less than 1.00 is ever significant. Write your findings here.

7. For the first *t* test, the authors state that $t = 0.23$, $p = 0.82$. If this is true, is it also true that $t = 0.23$, $p > .05$? Would both statements lead you to the same conclusion? Explain.

8. How many null hypotheses were tested in the excerpt? Explain.

Part B: Questions for Discussion

9. To compute a value of *t* for dependent groups, individuals in one group must be paired or matched in the other group. Was this true in the study described in the excerpt? Explain.

10. In the excerpt, the researchers state that the drug usage of the Supportive Counseling subjects "was unchanged" from pre-treatment to end of treatment. Yet, Figure 1 clearly shows a change. Speculate on the reason for this apparent discrepancy.

11. Near the end of the article from which the excerpt was drawn, the researchers state that the two significant differences for the Behavioral Counseling group were not only "statistically significant" but also "clinically significant." Speculate on how "clinical significance" differs from "statistical significance."

Exercise 31   Correlates of Fathering Behaviors

Significance of a Correlation Coefficient: I

Statistical Guide

To review the Pearson r, which is the most popular correlation coefficient, see the statistical guide for Exercise 19.

For a correlation coefficient based on data for a sample randomly drawn from a population, the null hypothesis states that the observed value is a chance deviation from a true value of 0.00 in the population (i.e., the observed value of r is not truly different from 0.00). When the probability that the null hypothesis is true equals 5 or less in 100, most researchers reject it and declare the coefficient to be statistically significant (i.e., reliable). The lower the probability, the more significant the relationship.

Excerpt from the Research Article[1]

Children were between the ages of 9 and 12, with a mean age of 10.8 years.

Parental involvement. The Parent Involvement Scale consisted of 12 items focusing on different areas of potential parent-child activity, such as "Celebrating holidays with child," [and] "Attending school or church related functions"....

Parental behavior. The Children's Report of Parental Behavior [yields scores], with a higher score representing a higher frequency of supportive behaviors and a lower frequency of critical or angry behaviors.

Parenting together. The Parenting Together Scale...consisted of 11 items describing possible aspects of partners' parenting collaboration (e.g., "Discussing personal problems the child may be experiencing").

Self-concept. The Self-Concept scale...consists of 15 items related to children's feelings of self-worth ... with a higher score indicating a more positive self-concept.

Psychological problems. Each parent responded to a 19-item psychological and behavioral problem scale...(e.g., "Restless, jumpy, hyperactive," "Bullying or mean to animals or other children," "Has trouble sleeping").

Classroom behavior. The instrument's total score...represented teachers' overall impression of adjustment, as reflected in classroom behavior, with a higher score indicating more positive adjustment.

Peer popularity. [The] Peer Friendship Survey asked all children, anonymously, to indicate which classmates, if any, were their best friends.

---

[1] Source: Bronstein, P., Stoll, M. F., Clauson, J., Abrams, C. L., & Briones, M. (1994). Fathering after separation or divorce: Factors predicting children's adjustment. *Family Relations*, *43*, 469–479. Copyrighted 1994 by the National Council on Family Relations, 3989 Central Ave. NE, Suite 550, Minneapolis, MN 55421. Reprinted by permission.

Table 2 *Correlations of parenting measures with child adjustment measures [for fathers in two-biological parent households (n = 79)]*

| | Child Adjustment | | | | | |
Measure	Self-Concept	Psych. Problems	GPA	Classroom Behaviors	Peer Popularity	Family Income
Family Income	.34***	−.08	.40***	.27**	.11	
Fathers' Parenting						
Involvement	.41***	.02	.29**	.20†	.06	.29**
Parental Behavior	.57***	−.02	.30**	.14	.05	.32**
Parenting Together	.18†	−.36***	.24*	.39***	−.05	.25***

†$p < .1$. *$p < .05$. **$p < .01$. ***$p < .001$. [All tests are one-tailed.]

Questions for Exercise 31

Part A: Factual Questions

1. Which two variables have the strongest relationship between them?

2. The correlation between family income and GPA indicates that those who are lower on income tend to be

   A. lower on GPA.   B. higher on GPA.

3. Is the relationship between family income and involvement direct *or* inverse?

4. Is the relationship between parenting behavior and psychological problems direct *or* inverse?

5. Is the relationship between involvement and psychological problems statistically significant? Explain.

6. Should the null hypothesis for the relationship referred to in question 5 be rejected?

7. All the correlations with GPA are significant. Which one is significant at the highest level? Explain.

8. What is the probability that the value of .34 (the *r* in the upper-left corner of the table) is a random deviation from a true correlation of 0.00?

9. The value of *r* for the relationship between peer popularity and parental behavior is .05. Does this mean that the correlation coefficient is statistically significant at the .05 level? Explain.

10. The value "$p < .001$" in the footnote to the table indicates that the probability that the null hypothesis is true is less than one in

    A 10.   B. 100.   C. 1,000.

11. Write out in words, without using numerals, the meaning of "$p < .1$," which appears in the footnote to the table.

Part B: Questions for Discussion

12. The footnotes to the table indicate that two of the coefficients have a *p* value of .1 associated with them. If you have a statistics textbook, consult it to see if the author recommends using this level when testing for significance.

13. The footnote to the table indicates that all tests are one-tailed. If you have a statistics textbook, look up the discussion of one-tailed and two-tailed tests. For a given set of data, which type of test is more likely to lead to rejection of the null hypothesis?

14. Describe in words, without using numbers, the strength and direction of the relationship between parenting together and psychological problems.

15. Describe in words, without using numbers, the strength and direction of the relationship between parental behavior and psychological problems.

Exercise 32   Workaholism
Significance of a Correlation Coefficient: II

Statistical Guide

To review the Pearson *r*, see the statistical guide for Exercise 19. To review the coefficient of determination, see the statistical guide for Exercise 20. To review the meaning of the significance of *r*, see the statistical guide for Exercise 31. Keep in mind that the lower the probability, the higher the level of significance.

Background Information

Of the entire population of 4,000 employees in a high technology company, a random sample of 503 were invited to participate in the study. Of these, 175 (35%) employees (74% men and 82% married) participated. The excerpt shown below shows a sample item from each instrument on which Table 1 is based.

Excerpt from the Research Article[1]

Potential participants (*N* = 503) were randomly selected from a database for the entire population of salaried employees…in a large, high technology corporation located in the U.S. [They were administered the following scales.]

Driven scale sample item: "I feel guilty when I take time off from work."

Work Involvement sample item: "Between my job and other activities I'm involved in, I don't have much free time."

Enjoyment of Work scale sample item: "My job is so interesting that it often doesn't seem like work."

Work-life Conflict sample item: "After work, I come home too tired to do some of the things I'd like to do."

Life Satisfaction sample item: "If I could live my life over, I would change almost nothing."

Purpose in Life sample item: "I am usually (1) *completely bored* to (7) *exuberant, enthusiastic*."

Table 1   *Summary statistics and intercorrelations for workaholism scales and outcome measures*

Measure	Driven	WI	Enjoy	WLC	LS	PIL	*M*	*SD*
Driven	—	.24**	.00	.42**	−.20**	−.08	18.91	3.79
WI		—	.08	.20**	−.20**	.03	17.78	4.46
Enjoy			—	−.14	.37**	.42**	20.68	5.92
WLC				—	−.24**	−.17*	13.05	3.47
LS					—	.53**	21.49	6.43
PIL						—	107.23	13.54

Note. N = 171. WI = Work Involvement scale; Enjoy = Enjoyment of Work scale; WLC = Work-life Conflict; LS = Life Satisfaction scale; PIL = Purpose in Life scale.
*p < .05, two-tailed. **p < .01, two-tailed.

---

[1] Source: Bonebright, C. A., Clay, D. L., & Ankenmann, R. D. (2000). The relationship of workaholism with work-life conflict, life satisfaction, and purpose in life. *Journal of Counseling Psychology, 47*, 469–477. Copyright © 2000 by the American Psychological Association, Inc. Reprinted with permission.

Questions for Exercise 32

Part A: Factual Questions

1. What is the correlation coefficient for the relationship between Life Satisfaction and Purpose in Life?

2. What are the names of the two scales between which there is the weakest correlation?

3. Is the correlation coefficient for the relationship between Enjoyment of Work and Purpose in Life statistically significant? If yes, at what probability level is it significant?

4. The relationship between WLC and LS is such that those who have high WLC scores tend to have:

   A. high LS scores.   B. low LS scores.

5. "Driven" has *in*significant relationships with which other variables? Identify them by their abbreviations.

6. There are 10 statistically significant correlation coefficients in the table. Are they all significant at the same level of significance? Explain.

7. Should the null hypothesis for the relationship between WI and LS be rejected? Explain.

8. What is the probability that the correlation for the relationship between PIL and Enjoy is a chance deviation (due to random sampling error) from 0.00?

9. In statistics, a Type I error is the error of rejecting the null hypothesis when, in truth, it is correct. If we reject the null hypothesis regarding the relationship between Driven and WI, what is the probability that we are making a Type I error?

10. How many of the correlations with PIL are statistically significant?

Part B: Questions for Discussion

11. Does it surprise you that some of the negative correlation coefficients are statistically significant? Explain.

12. The correlation between WI and LS is –.20, which in this study is statistically significant. Would you characterize this relationship as being "very strong"? Explain.

13. Are the values in the column labeled "*M*" correlation coefficients? Explain.

14. Does it surprise you that the relationship between Life Satisfaction and Purpose in Life is statistically significant? Explain.

Exercise 33 Extracurricular Activities and Dropping Out
One-Way ANOVA

Statistical Guide

Like a t test, a one-way analysis of variance (ANOVA) can be used to determine the significance of the difference between two means. Instead of yielding a value of t, however, an ANOVA yields a value of F. For a given set of data, t and F will yield the same probability (p) that the null hypothesis is true. Thus, the two tests are interchangeable when comparing two means. As with the t test, when an ANOVA yields a small value of p (such as .05, .01 or .001), the null hypothesis is rejected. A one-way ANOVA is also known as a univariate ANOVA.

Excerpt from the Research Article[1]

This study examined the relation between involvement in school-based extracurricular activities and early school dropout. Longitudinal assessments were completed for 392 adolescents (206 girls, 186 boys).... *Early school dropout* was defined as failure to complete the 11th grade.

To evaluate whether extracurricular involvement would predict early school dropout, we compared activity participation across grades 7 to 10 for dropouts and nondropouts. Univariate ANOVAs were performed separately at each grade.... Figure 2 shows the mean number of activities participated in by nondropouts and dropouts. Dropouts participated in significantly fewer extracurricular activities at all grades, even several years prior to dropout; 7th grade, $F(1, 389) = 8.41$, $p < .01$; 8th grade, $F(1,365) = 10.14$, $p < .001$; 9th grade, $F(1, 343) = 15.46$, $p < .001$; and 10th grade, $F(1, 314) = 31.00$, $p < .001$.

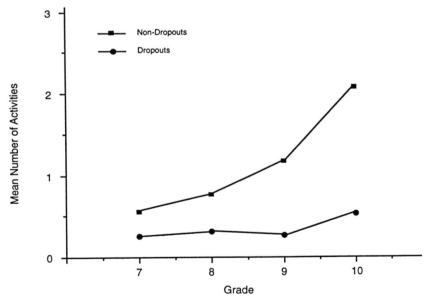

Figure 2 *Mean number of extracurricular activities participated in as a function of dropout status and grade*

---

[1] Source: Mahoney, J. L., & Cairns, R. B. (1997). Do extracurricular activities protect against early school dropout? *Developmental Psychology, 33*, 241–253. Copyright © 1997 by the American Psychological Association, Inc. Reprinted by permission.

Questions for Exercise 33

Part A: Factual Questions

1. According to Figure 2, at which grade level was the average number of extracurricular activities for dropouts the most similar to the average number for nondropouts?

2. Is the difference referred to in question 1 statistically significant? If yes, at what probability level is it significant?

3. Should the null hypothesis be rejected for the difference referred to in question 1? If so, at what probability level should it be rejected?

4. What is the probability that random sampling errors created the mean difference between the two groups at the 8th grade level?

5. Should the null hypothesis be rejected for the difference between the two means at the 9th grade level? Explain.

6. How many means are being compared at each grade level?

7. The difference at the 7th grade is

   A. more statistically significant than the differences at the other grade levels.
   B. less statistically significant than the differences at the other grade levels.

Part B: Questions for Discussion

8. The researchers indicate that their study was *longitudinal*. What do you think this term means?

9. The researchers do not report the standard deviations associated with the means. Would you be interested in knowing the standard deviations? Explain.

10. If you have studied how to compute an ANOVA, explain the meaning of the values in parentheses immediately after each F. In other words, what is the name for these values, and on what are they based?

11. In your opinion, have the researchers established that participation in extracurricular activities *causes* a reduction in dropping out? Explain.

Exercise 34   Gender and Age Differences in Fears

One-Way ANOVA with Confidence Intervals

Statistical Guide

To review the one-way ANOVA, see the statistical guide for Exercise 33. To review confidence intervals for means, see the statistical guides for Exercises 25 and 26. Note that 95% confidence intervals are reported in the excerpt. To review the median, see the statistical guide for Exercise 9. Also note that a value of F of less than 1 in an ANOVA is never significant. Also note that in a table where low probabilities are indicated by footnotes to the table, the absence of a footnote for a value of F indicates that the associated difference is *not* significant.

Excerpt from the Research Article[1]

A questionnaire was mailed together with a stamped return envelope [to 1,000 subjects]. Subjects were asked to complete...ratings of [their fears]. Zero denoted "no fear at all" and 100 "maximal fear."

To define relatively younger and older Ss, a median (Md = 41.0) split resulted in an average ($\pm$SD) age of 29.0 yr ($\pm$6.5) in the younger and of 53.3 yr ($\pm$8.4) in the older group.

Table 3   *Average ($\pm$SD) ratings of fear and confidence intervals (CIs). Gender and age differences are tested by analysis of variance.*

	Men	CIs	Women	CIs	
Gender Differences					
Animal fear					
Snakes	22.3 (24.4)	19.6–25.1	38.1 (31.5)	34.8–41.4	$F(1,657) = 50.8$***
Spiders	12.9 (19.0)	10.8–15.1	27.2 (29.6)	24.1–30.3	$F(1,659) = 53.1$***
Mutilation fear					
Injections	17.5 (24.9)	14.7–20.3	21.7 (26.9)	18.9–24.5	$F(1,659) = 4.4$*
Dentists	17.9 (25.9)	15.0–21.0	23.6 (26.7)	20.8–26.4	$F(1,658) = 7.9$**
Injuries	23.4 (24.8)	20.6–26.2	32.4 (29.5)	29.3–35.5	$F(1,660) = 17.8$***

	Younger	CIs	Older	CIs	
Age Differences					
Animal fear					
Snakes	31.0 (29.0)	27.8–34.2	31.0 (29.9)	27.7–34.2	$F(1,651) = <1$
Spiders	25.2 (28.7)	22.0–28.4	16.2 (22.6)	13.8–18.6	$F(1,653) = 19.8$***
Mutilation fear					
Injections	23.0 (28.4)	19.9–26.1	16.9 (23.4)	14.4–19.4	$F(1,653) = 9.0$**
Dentists	21.8 (27.7)	18.7–24.9	20.4 (25.4)	17.7–23.1	$F(1,652) < 1$
Injuries	26.6 (27.2)	23.6–29.6	29.9 (28.2)	26.8–32.9	$F(1,654) = 2.2$

*$p < 0.05$; **$p < 0.01$; ***$p < 0.001$

---

[1] Reprinted from: Fredrikson, M., Annas, P., Fischer, H., & Wik, G. (1996). Gender and age differences in the prevalence of specific fears and phobias. *Behaviour Research and Therapy*, *34*, 33–39 with kind permission from Elsevier Science Ltd., The Boulevard, Langford Lane, Kidlington, 0X5 1GB, UK. Copyright © 1996 by Elsevier Science, Ltd.

Questions for Exercise 34

Part A: Factual Questions

1. On the average, were men or women more fearful of snakes? Explain the basis for your answer.

2. We can be 95% confident that the average rating on fear of snakes for men in the population is between what two values?

3. Do the confidence intervals on fear of snakes for men and women overlap? (That is, are some of the values in the interval for men also included in the interval for women and vice versa?)

4. Suppose that, based on information in Table 3, a researcher said, "Odds are very high that the true rating by the population of younger adults on fear of dentists is somewhere between 18.7 and 24.9, while odds are very high that it is somewhere between 17.7 and 23.1 for older adults." On the basis of this statement, would it be reasonable to assert that older and younger adults are truly different in their fear of dentists? Explain.

5. Is the difference between older and younger adults on fear of dentists statistically significant? Explain how you got the answer. Is your answer to this question consistent with your answer to question 4?

6. Should the null hypothesis for the difference between older and younger adults on fear of injuries be rejected? Explain the basis for your answer.

7. How many of the values of *F* were *not* statistically significant?

8. According to Table 3, which of the following statements about the difference between younger and older adults on fear of injections is true?
   A. It is unlikely that sampling error created the difference between the means.
   B. It is likely that sampling error created the difference between the means.

9. Which of the three gender differences in mutilation fear is statistically significant at the highest level? Explain.

10. In light of the confidence intervals, which of the five types of fears is greatest among women? Explain.

Part B: Questions for Discussion

11. Suppose you had conducted this study and examined the means, standard deviations, and confidence intervals for younger and older adults on fear of dentists before conducting an ANOVA. Would you have expected to find a significant difference between the two means? Explain.

12. The values "1,657" in the first row of values in Table 3 are degrees of freedom. If you have studied how to conduct an ANOVA, you should be able to determine from this whether all 1,000 people who were mailed questionnaires answered this question. Did they? Explain.

13. In your opinion, is it possible that men are, on the average, just as fearful as women but are less likely to admit to their fear than women are? If yes, how likely do you think it is that this possibility distorted the results?

14. The title of Table 3 states that the table contains "Average (±*SD*) ratings...." Which average (mean, median, or mode) is probably being reported? Explain the basis for your answer.

15. Explain what a "median split" is and how it was used in this study.

Exercise 35   Effectiveness of Assertiveness Training
Two-Way ANOVA: I

Statistical Guide

To review the general purpose of ANOVA, see the statistical guide for Exercise 33. In a two-way ANOVA, there are two independent variables (usually nominal classification variables) and one outcome variable (usually a continuous score variable). To understand this, examine Figure 1 in the excerpt, where time of testing (pretest and posttest) is one of the classification variables, and level of nursing (nursing assistants, licensed vocational nurses, and registered nurses) is the other one. The outcome variable is discomfort scores.

A two-way ANOVA allows us to examine interactions between the independent variables. For example, if the different types of nurses responded differently to the treatments given between the pretest and posttest, we would say that there is an interaction (i.e., the treatments interacted with level of nursing skill).

Excerpt from the Research Article[1]

The participants in this study were the 62 members of the nursing staff working in a 47-bed Department of Veterans Affairs spinal cord injury center.

The purpose of this study was to evaluate the effectiveness of the training course in [reducing discomfort when engaging in difficult] staff interactions. A behavioral approach was selected as the basis for the training. It consisted of a combination of techniques such as lecture and discussion, modeling, and behavior rehearsal, with feedback provided by group members and the trainers….

The study included the Spinal Cord Injury Assertiveness Inventory (SCIAI) as preprogram and postprogram measures to assess the effectiveness of the training…. At the request of the nurse managers, all nursing staff participated in the training, so we had no control group.

A repeated measures ANOVA on the total score of the SCIAI showed no simple effect [i.e., main effect, $p = .1069$] related to the time of testing [i.e., the mean on the pretest ($M = 2.40$) for all participants was not significantly different from the mean on the posttest for all participants ($M = 2.27$)]…. However, when we examined the participants' education, we found a statistically significant interaction effect between the time of testing and education ($F_{2,58} = 3.468$, $p = .0378$) (see Figure 1). After the class, the NAs (nursing assistants), all of whom had a high school education, showed an increased discomfort level, whereas the discomfort level of the LVNs (licensed vocational nurses) and the RNs (registered nurses) decreased. However, paired t tests on the various groups' data showed that only the RNs ($t = 2.692$, $df = 36$, $p = .0107$) exhibited a statistically significant change in their discomfort level related to time of testing.

---

[1] Source: Dunn, M., & Sommer, N. (1997). Managing difficult staff interactions: Effectiveness of assertiveness training for SCI nursing staff. *Rehabilitation Nursing, 22*, 82–87. Reprinted from *Rehabilitation Nursing, 22* (2), 82–87 with permission of the Association of Rehabilitation Nurses, 4700 West Lake Ave., Glenview, IL 60025-1485. Copyright © 1997.

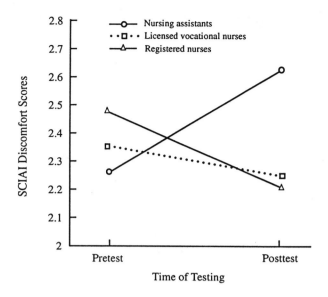

Figure 1 *Interaction between time of testing (pretest vs. posttest) and educational level*

Questions for Exercise 35

Part A: Factual Questions

1. Should the null hypothesis for the main effect (for time of testing) be rejected? Explain.

2. Does Figure 1 indicate that there is an interaction? Explain.

3. Is the interaction statistically significant at the .05 level?

4. Is the interaction statistically significant at the .01 level?

5. What is the precise probability that the null hypothesis regarding the interaction is true?

6. Did the discomfort of the nursing assistants change significantly from pretest to posttest? Explain.

7. Should the null hypothesis regarding the change in the registered nurses' level of discomfort be rejected? Explain.

Part B: Questions for Discussion

8. If the findings are correct, would it be desirable to use the training with all types of nurses in the future? Explain.

9. The authors could not use a control group. Is this a limitation of the study? Explain.

10. Suppose you conducted a similar study but used a different type of training that produced a reduction in discomfort scores equally strong in all three groups of nurses. Draw a figure like the one in the excerpt showing this hypothetical result.

Exercise 36   Aggression, Gender, and Ethnicity
Two-Way ANOVA: II

Statistical Guide

To review the general purpose of ANOVA, see the statistical guide for Exercise 33. To review the more specific purpose of a two-way ANOVA, see Exercise 35.

In the excerpt, the main effect for gender (as indicated by F_{gender}) is based on a comparison of the mean for all women and the mean for all men. The main effect for ethnicity (as indicated by $F_{ethnicity}$) is based on a comparison of the mean for all Anglos and the mean for all Hispanics.

Excerpt from the Research Article[1]

Subjects were 259 female and 102 male students (plus two who did not indicate their gender) who were enrolled in 11 undergraduate and 2 graduate classes at a large urban university…with a median age of 24.

The instrument was an anonymous questionnaire [that included] a list of 18 experiences and asked [the subjects] to indicate how many times they had "ever had each of the following experiences." Responses were coded as 0 (*if it has never happened to you*), 1 (*if it has happened to you once or twice*), 2 (*3 to 5 times*), 3 (*6 to 10 times*), and 4 (*more than 10 times*).

Table 1 presents the mean scores of male and female Anglos and Hispanics in their responses to the questions about the frequency with which they had experienced each of the 18 types of aggression…. None of the interactions between ethnicity and gender approached significance.

Table 1   *Mean scores for Anglo and Hispanic males and females on experiences as target*

	Males		Females			
Item	Anglo	Hisp	Anglo	Hisp	F^a_{gender}	$F^a_{ethnicity}$
Something thrown at you	2.54	2.05	1.38	1.15	48.15***	3.79
Pushed, grabbed, shoved	2.89	2.62	2.09	1.73	24.07***	4.08*
Slapped	1.67	1.29	1.19	0.94	7.15**	3.13
Kicked, bitten, hit by fist	2.46	2.38	1.14	0.89	60.68***	1.47
Hit with object	2.18	1.95	0.93	0.85	52.52***	0.48
Beaten up	0.97	1.05	0.34	0.40	24.11***	0.27
Threatened with gun or knife	0.67	0.62	0.22	0.13	28.57***	0.91
Knife or gun used against you	0.38	0.24	0.06	0.05	19.84***	0.59
Teased meanly	2.83	2.48	2.63	1.90	2.83	12.28***
Yelled at	3.25	2.95	3.34	2.95	0.16	5.52*
Called cruel, unethical, dishonest	1.08	1.24	0.78	0.75	5.39*	0.03
Called stupid, worthless	1.75	1.00	1.59	1.38	0.01	3.32
Called an obscene name	2.92	2.76	2.32	1.84	13.85***	4.82*
Received an obscene gesture	3.10	2.57	2.45	1.92	14.69***	9.91**
Forced to have sex	0.07	0.05	0.58	0.25	16.70***	5.65*
Treated condescendingly	2.20	1.76	2.51	1.38	0.45	25.55***
Honked at loudly	2.44	2.10	2.50	1.65	0.19	17.85***
Race, culture insulted	1.58	2.05	0.96	0.93	20.37***	0.42

[a]Degrees of freedom range from 1 and 292 to 1 and 299.

*$p < .05$. **$p < .01$. ***$p < .001$.

---

[1] Reprinted with permission from: Harris, M. B. (1996). Aggressive experiences and aggressiveness: Relationship to ethnicity, gender, and age. *Journal of Applied Social Psychology, 26,* 843–870. © V. H. Winston & Son, Inc., 360 South Ocean Boulevard, Palm Beach, FL 33480. All rights reserved.

Questions for Exercise 36

Part A: Factual Questions

1. The average male Hispanic had something thrown at him about how many times?

2. Was there a significant difference on "having something thrown at you" between the two ethnic groups? Explain.

3. How many means were compared by the ANOVA to get the F value of 24.07 for "pushed, grabbed, shoved"?

4. How many means were compared by the ANOVA to get the F value of 4.08 for "pushed, grabbed, shoved"?

5. Overall, did the average male or average female experience having his or her race/culture insulted more often? Explain.

6. Should the null hypothesis be rejected for the difference referred to in question 5?

7. The average Anglo reported (a) a greater incidence of being teased meanly and (b) a greater incidence of being yelled at than the average Hispanic. Which of these two differences is significant at a higher level? Explain.

8. Were more of the gender differences or more of the ethnic differences statistically significant? Explain.

9. In all, how many null hypotheses should be rejected based on the information in Table 1?

Part B: Questions for Discussion

10. Why do you think the researcher did not give the *p* values for the interactions?

11. Examine the means for the first item in Table 1. Rewrite them in this table:

	Anglo	Hispanic
Males	$M =$	$M =$
Females	$M =$	$M =$

Do you prefer this type of table or Table 1 in the excerpt for reporting the means? Explain.

12. Is it possible to determine from the excerpt (including Table 1) whether the difference between the means for Anglo males and Hispanic males on "treated condescendingly" is statistically significant? Explain.

13. Would you be willing to generalize the results to Hispanics and Anglos enrolled in other large urban universities? Explain.

14. Are the results interesting to you? Explain.

Exercise 37   Admitting to Gun Ownership

Standard Error of a Percentage and 95% Confidence Interval

Statistical Guide

To review the percentages, see the statistical guide for Exercise 1. The standard error of a percentage is a margin of error (i.e., an allowance for error that we use when estimating the population percentage from a sample drawn at random from a population). When confidence intervals are reported, they are often at the 95% level. For such an interval, chances are 95 in 100 that the true percentage in the population lies within the interval.

In the following excerpt, the researchers give the 95% margins of error for the associated percentage in parentheses in the table. To get the limits of a 95% confidence interval (95% CI), simply add the margin to the mean and then subtract it from the mean, as indicated by "±" in the parentheses.

Excerpt from the Research Article[1]

Since easy access to firearms is considered to be a risk factor for suicides, homicides in the home, and unintentional firearm-related injury and death, data on the presence of guns in the home are needed to guide public policy.

Random samples were selected from two populations in Ingham County [Michigan] thought likely to have guns in their households—(a) people who had purchased a hunting license and (b) people who had registered a handgun. [Telephone interviews were conducted within a year of these activities.]

The question—"Some people keep guns in their household. Are guns of any kind kept in your household?"—[was asked.]

If it can be assumed that all of the people who purchased a hunting license owned a gun and that all of the persons who registered a handgun were still in possession of the handgun and kept it in their household, then 11.4 percent of the responses were invalid and would result in underreporting of household gun presence.

Table 2   *Percentages and 95% confidence interval (CI) limits of respondents from likely gun households who reported the presence of a gun in the home [that is, answered "yes" to the question][1]*

Samples	Total	95% CI	18–34	95% CI	35–54	95% CI	55 and over	95% CI
				Age (in years)[2]				
Likely gun households[3]	88.6	(±4.7)	79.7	(±10.3)	94.1	(±5.0)	90.6	(±10.2)
Hunting licensees	89.7	(±6.1)	84.4	(±12.6)	93.5	(±7.2)	89.5	(±13.8)
Handgun registrants	87.3	(±7.3)	74.1	(±16.6)	94.9	(±6.9)	92.3	(±15.5)

[1]Respondents who refused to answer the question were not included in this analysis.
[2]Age of respondent reflects the age of the oldest person in the household of the same sex as the registrant or licensee.
[3]Combined sample of handgun registrant households and hunting licensee households.

[1] Source: Rafferty, A. P., Thrush, J. C., Smith, P. K., & McGee, H. B. (1995). Validity of a household gun question in a telephone survey. *Public Health Reports, 110*, 282–288.

Questions for Exercise 37

Part A: Factual Questions

1. Of likely gun households, which age group had the smallest percentage answering "yes" to the question?

2. For the group referred to in question 1, what percentage answered "no"?

3. Consider all the percentages in Table 2. Which group in which sample had the lowest percentage reporting "yes"?

4. For the group referred to in question 3, what is the value of the margin of error for the percentage?

5. For the group referred to in question 3, what are the limits of the 95% confidence interval for the percentage?

6. We can have 95% confidence that the true percentage of all hunting licensees in the population who would answer "yes" is between which two values?

7. For likely gun households, the 95% confidence interval for the 18- to 34-year-olds is 69.4 and 90.0. Does this interval overlap with the 95% confidence interval for the 35- to 54-year-olds in likely gun households (that is, are any of the values in the latter interval included in the interval from 69.4 to 90.0)?

8. Confidence intervals are estimates of the true values in populations. When the intervals for two groups on the same question overlap, we should conclude that we have failed to establish a reliable difference. In light of this principle, should we conclude that there is a reliable difference between hunting licensees in the 35- to 54-year-old group and handgun registrants in the same age group?

9. For the total sample, is there a reliable difference at the 95% level between handgun registrants and hunting licensees? Explain.

Part B: Questions for Discussion

10. Consider the following statements about the results for the total group of likely gun households. If a researcher could report only one of these statements, which one would you recommend that she or he report? Note that the values in choice B are the limits of the 95% confidence intervals. Explain the basis for your answer.

   A. "Based on our survey, we estimate that about 88.6% of the total population of likely gun households would report the presence of a gun in the home when questioned by telephone."
   B. "Based on our survey, we have 95% confidence that between 83.9% and 93.3% of the total population of likely gun households would report the presence of a gun in the home when questioned by telephone."

11. The first footnote indicates that those who refused to answer the question were (necessarily) omitted from the analysis. Could the omission of the refusers cause a bias when trying to estimate the prevalence of guns in households? Explain.

12. To protect the privacy of the participants, those who conducted the interviews knew only the sex and telephone number of the registrants and licensees. Thus, they were unable to ask to speak with the registrant or licensee. Instead, when they called, they asked to speak with the oldest person in the household who was of the same sex (for example, "May I speak with the oldest male in your household?"). This is indicated in the second footnote in the table. Is this a limitation of the study? Explain.

13. In their article, the researchers note that there was "noncoverage of households that do not have telephones, which is a limitation…." Do you agree that it is a limitation? Explain.

Exercise 38   Cutting Class and Other Behaviors
Chi Square: I

Statistical Guide

Chi square is a test of the significance of the differences among frequencies. Its symbol is χ^2. The null hypothesis for a chi square test asserts that the differences were created by random sampling error. Like other significance tests, the null hypothesis is rejected when the probability that it is true is low—such as .05, .01, .001, or less. The lower the probability, the more significant the differences. Note that if the frequencies are significantly different, the percentages based on them are also significantly different. Also note that in tables that present the results of chi square tests, it is customary to footnote only those differences that were declared to be statistically significant.

Excerpt from the Research Article[1]

We surveyed 120 sophomores and juniors enrolled in general education psychology courses.... All were between the ages of 18 and 23. Students were surveyed in their classes by an anonymous self-report instrument. We determined absenteeism [cutting class] from a single item, "I have missed a class during the past month for no valid reason," and we compared this item to reports of nine other behaviors, also in a yes/no format as to whether they had occurred in the last month. Seven of the behaviors were problematic behaviors (broken the speed limit, slapped or hit someone...), and two were positive behaviors (visiting family members and reading a book not assigned in class). We also included two status variables, "I have a tattoo" and "I have a body part pierced other than my ears."

Table 1   *Number and percentage of students answering yes to other behaviors in the groups reporting class cutting (n = 68) and not cutting (n = 52)*

Behavior	Cutting		Not cutting		χ^2
	n	%	*n*	%	
Got drunk	59	87	24	46	22.79**
Sped	63	93	39	75	7.19*
Broke the law	35	51	10	19	13.07**
Told a significant lie	14	21	8	15	0.53
Thought about dropping out	8	12	3	6	0.79
Was depressed	7	10	5	10	0.02
Hit/slapped	8	12	11	21	1.95
Tattoo	13	19	4	8	3.16
Body pierced	18	26	7	13	3.17
Read a book not for class	25	37	15	29	0.83
Visited family	62	91	40	77	4.61*

* *p* < .05. ** *p* < .002.

---

[1] Source: Trice, A. D., Holland, S. A., & Gagné, P. E. (2000). Voluntary class absences and other behaviors in college students: An exploratory analysis. *Psychological Reports, 87,* 179–182. Copyright © 2000 by Psychological Reports. Reprinted with permission.

Questions for Exercise 38

Part A: Factual Questions

1. In the last column of Table 1, the value of chi square (χ^2) is shown to be 0.02 for "was depressed." Does the 0.02 indicate that the difference is statistically significant? Explain.

2. What percentage of those who did not cut class reported reading a book not for class?

3. For "thought about dropping out of school," what is the percentage difference between those who cut and those who did not cut? (Note: Subtract the smaller difference from the larger difference.)

4. Is the difference for "sped" statistically significant? If yes, at what level?

5. For "was depressed," is the difference between those who cut and those who did not cut statistically significant? What is the basis for your answer?

6. Should the null hypothesis for the difference between the two groups for "got drunk" be rejected? Explain.

7. How many of the null hypotheses were rejected? Explain the basis for your answer.

8. For which of the following behaviors is the difference more significant: "sped" *or* "broke the law"? Explain the basis for your answer.

9. Table 1 indicates that the difference for "broke the law" is statistically significant at the .02 level. Is it also statistically significant at the .05 level? Explain.

Part B: Questions for Discussion

10. Do you agree with the researchers that getting a tattoo and piercing the body are "status variables"? Explain.

11. The researchers measured the variables in a yes/no format. In your opinion, would it also be of interest to measure frequency (e.g., how many times participants got drunk) and intensity (how drunk did they get)?

12. The researchers reported significance at the .05 and .002 levels. If you have a statistics textbook, examine it to see which probability levels are mentioned when interpreting a chi square test, and write your findings here.

13. Are larger or smaller values of chi square in Table 1 statistically significant?

14. What is your opinion on using a one-month period for this study? If you were doing a study on this topic, would you have used a one-month period? A shorter period (e.g., tattooed in the last week) or longer period (e.g., tattooed in the last year)? Explain.

Exercise 39   Status of Women Inmates

Chi Square: II

Statistical Guide

To review chi square, see the statistical guide for Exercise 38.

Excerpt from the Research Article[1]

Subjects were inmates incarcerated in Louisiana's only prison for adult female offenders. The crimes for which the inmates were serving time were dichotomized into violent or nonviolent. Violent crimes included murder, manslaughter, negligent homicide, battery, robbery, attempted robbery, and child abuse.

Table 2 lists the significant relationship between marital status and criminal variables. Marital status was significant for the type of crime committed ($\chi^2 = 24.27$, $p \le .0001$).

Table 2 [also] indicates the significant relationship between violence and marital status ($\chi^2 = 6.31$, $p \le .05$). Single subjects were least likely to have committed violent crimes; married subjects were most likely to have committed violent crimes. Previously married subjects fell between these two extremes. This is opposite of the predicted relationship.

Table 3 indicates a significant association in number of inmates who were convicted of various types of crime by occupational status prior to arrest.... Table 3 [also] shows a significant ...relationship between violence and employment status.

Table 2   *Percentage of subjects for each marital status*

Variable	Marital Status						χ^2
	Single		Previously Married		Married		
	%	n	%	n	%	n	
Crime Type							
Murder	12.20	5	46.15	18	46.15	18	
Robbery	24.39	10	2.56	1	17.95	7	
Theft	53.66	22	28.21	11	23.08	9	
Drug & Other	9.76	4	23.08	9	12.82	5	24.27**
Violence							
No	63.64	28	53.49	23	36.59	15	
Yes	36.36	16	46.51	20	63.41	16	6.31*

*$p \le .05$ **$p \le .0001$

[1] Reproduced with permission of the authors and publisher from: Campbell, C. S., & Robinson, J. W. (1997). Family and employment status associated with women's criminal behavior. *Psychological Reports*, *80*, 307–314. Copyright © Psychological Reports 1997.

Table 3 *Percentage of subjects for each employment status*

Variable	Unemployed		Employed		χ^2
	%	*n*	%	*n*	
Crime Type					
Murder	60.87	14	29.49	23	
Robbery	13.04	3	12.82	10	
Theft	17.39	4	39.74	31	
Drug & other	8.70	2	17.95	14	8.30*
Violence					
Violent	69.23	18	42.68	35	
Nonviolent	30.77	8	57.32	47	5.57*

*$p \le .05$

Questions for Exercise 39

Part A: Factual Questions

1. Which marital status group had the highest percentage found guilty of theft?

2. The implied null hypothesis for the first significance test in the excerpt is that there are no *true* differences in the types of crime committed across marital status groups. Should this null hypothesis be rejected? Explain.

3. State the implied null hypothesis for the second significance test.

4. Which of the following relationships is statistically significant at a higher level? Explain the basis for your choice.

   A. Crime type and marital status   B. Violence and marital status

5. In Table 2, the chi square value of 6.31 has a footnote, which indicates that *p* is equal to *or* less than .05. If you have a statistics book, look up 6.31 with 2 degrees of freedom in the table of critical values. Is it significant at *exactly p* = .05? Explain.

6. Is the association between employment status and crime type statistically significant? Explain.

118

7. Describe in your own words the general nature of the association between employment status and violence. Do not refer to the specific values of the percentages in your statement.

8. Should the null hypothesis for the association you described in question 7 be rejected? Explain.

9. The association between employment status and violence is reported to be significant at the .05 level. Is it also significant at the .01 level? (Hint: You should be able to figure out the answer using logic. If not, refer to the table of critical values using $df = 1$.)

Part B: Questions for Discussion

10. Many researchers report the df associated with each chi square test. The researchers who wrote the excerpt did not. In your opinion, is this an important omission?

11. The researchers state in the excerpt that the association between violence and marital status was the opposite of what they had predicted (i.e., hypothesized). Before reading this excerpt, would you have predicted that married and previously married women would be more likely to commit violent crimes than single women? Explain.

12. Would you be willing to generalize the results in the excerpt to women offenders in states other than Louisiana (that is, would you be willing to assume that what is true of women offenders in Louisiana is also true of women offenders in other states)? Explain.

Exercise 40   Barriers to Birth Control Use

Chi Square: III

Statistical Guide

To review chi square, see the statistical guide for Exercise 38.

Excerpt from the Research Article[1]

This study examined the relationship between barriers to using birth control and actual use of birth control among a national sample of Mexican American adolescents. Participants were either over age 15 or sexually active (regardless of age).

Among males, significant differences were found on three items (it is/would be difficult to get a partner to use birth control with you; if you used birth control, your friends might think that you were looking for sex; it is difficult to get birth control).

The limitations of the present study should be noted. First, the accuracy of self-reported information can pose a serious problem.... Second, birth control methods were not differentiated, and there may be different barriers to using condoms versus the birth control pill, for example.

Table 1   *Birth control use at most recent sexual intercourse by barrier scale items (male)*

Barrier Scale Items	Used* % (n = 47)	Did Not Use % (n = 33)	p
Difficult to get partner to use birth control			
Agree or neutral	37	63	
Disagree	63	38	.038
Feel birth control interferes with pleasure			
Agree or neutral	40	45	
Disagree	60	55	.650
Friends might think you were looking for sex			
Agree or neutral	41	73	
Disagree	59	27	.007
It is difficult to get birth control			
Agree or neutral	33	73	
Disagree	67	27	.001
Using birth control is morally wrong			
Agree or neutral	50	67	
Disagree	50	33	.171
Birth control is too expensive			
Agree or neutral	28	34	
Disagree	72	66	.623
Birth control is a hassle to use			
Agree or neutral	28	42	
Disagree	72	58	.233
Birth control involves too much planning			
Agree or neutral	63	65	
Disagree	37	35	1.0

*Refers to males who used contraception at the most recent intercourse (versus those who did not).

---

[1] Source: Pesa, J. A., & Mathews, J. (2000). The relationship between barriers to birth control use and actual birth control use among Mexican American adolescents. *Adolescence, 35*, 695–707. Copyright © 2000 by Libra Publishers, Inc. Reprinted with permission.

Questions for Exercise 40

Part A: Factual Questions

1. How many males reported that they did not use contraception at the most recent intercourse?

2. What percentage of the males who used contraception reported that it was difficult to get their partners to use birth control? What is the corresponding percentage for males who did not use contraception?

3. Table 1 reports the results of how many chi square tests?

4. How many of the chi square tests are significant at the .05 level?

5. Consider the test for "birth control is too expensive." Is this relationship statistically significant at conventional levels? Explain.

6. Is there a statistically significant relationship between the "use variable" (i.e., used versus did not use) and the variable represented by the statement that birth control is a hassle to use? Explain.

7. The third item is significant at the .01 level because its value is .007. Is it also significant at the .05 level? Explain.

8. Is the chi square being used to test the differences between means (the most common average) in Table 1? Explain.

9. Should the null hypothesis for "use" (used/not used) and feeling that birth control interferes with pleasure be rejected? Explain.

Part B: Questions for Discussion

10. While the values of *p* are given, the values of chi square that led to the *p*-values are not given. In your opinion, is this a serious omission? Explain.

11. In your opinion, how serious is the first limitation mentioned by the researchers? Is there anything that can be done to increase the accuracy of the reports?

12. Do you agree that not differentiating between birth control methods is a limitation? If yes, how serious is it?

13. In the same study, females were asked the same questions. Significant differences were found for only the last two items (i.e., hassle to use and too much planning). Does it surprise you that the results were different for females than they were for males? Explain.

Exercise 41 Depression and Nonverbal Cues

Mann-Whitney *U* Test

Statistical Guide

The Mann-Whitney *U* test is a test of statistical significance for use with two independent groups. Unlike the *t* test or ANOVA, this test is appropriate when the data are naturally ordinal or when the data have been reduced from interval to ordinal because one or both of the distributions are highly skewed.

Like other significance tests, when the Mann-Whitney *U* test indicates that *p* is equal to or less than .05, we usually declare the difference to be statistically significant (i.e., reject the null hypothesis). The smaller the value of *p*, the more significant the difference.

Excerpt from the Research Article[1]

Forty-one consecutive male patients seeking outpatient treatment volunteered to participate in this study prior to receiving treatment. Thirty-one met…criteria for major depression…and had not yet received antidepressant therapy. [Of these, 25 were selected at random for this study.] Twenty-five [nondepressed] controls were obtained by advertising.

The range of ages was 20 to 34 years for depressed volunteers and 21 to 34 for controls. The mean ages were, respectively, 27.2 and 27.6 years. The range of education was 12 to 16 years for both groups. All participants possessed high school diplomas….

Videotapes of five male medical interns were used as sources of nonverbal cues…. In these tapes, the interns were engaged in gambling trials in which they attempted to earn variable amounts of money while playing at a slot machine. The amount of money at risk was determined by random sequence and could be either a nickel, quarter, or dollar. During the playing time but before payoff, the interns' facial expressions had been covertly videotaped for five trials.

…the depressed patients and the control subjects viewed the original tapes of the five medical interns for five gambling trials each. The subjects were directed by prerecorded instructions to guess the amount of money at risk using only the nonverbal cues presented.

There was a significant difference between the depressed group and the control group in ability to detect these particular nonverbal cues…. A Mann-Whitney *U* test assessing the difference between the medians of the two groups was performed. The median number of correct responses for the control group was 11 out of 25 tries and for the depressed group, the median was 5.5 correct responses out of 25 tries. (Chance = 8 correct responses.) The results are clearly significant ($U = 35.5, p < .002$).

This study indicates that untreated depressed men may have significantly impaired ability to interpret nonverbal facial cues.

---

[1] Source: Reproduced with permission of the authors and publisher from: Giannini, A. J., Folts, D. J., Melemis, S. M., Giannini, M. C., & Loiselle, R. H. (1995). Depressed men's lower ability to interpret nonverbal cues: A preliminary study. *Perceptual and Motor Skills*, *81*, 555–560. Copyright © Perceptual and Motor Skills 1995.

Questions for Exercise 41

Part A: Factual Questions

1. What is the average age of the depressed volunteers?

2. There were 5 interns taped for 5 trials each. While watching them, each participant had a choice of 3 amounts of money. Explain how this information could be used to calculate the chance level of 8 correct responses reported in the excerpt.

3. What was the average number of correct responses for the control group?

4. By how many points did the average of the control group exceed the average of the depressed group?

5. Was the difference referred to in question 4 statistically significant at the .001 level?

6. Was the difference referred to in question 4 statistically significant at the .01 level?

7. Should the null hypothesis for the difference referred to in question 4 be rejected at the .01 level?

Part B: Questions for Discussion

8. Speculate on why the researchers used the *mean* as the average for age but the *median* as the average for number of correct responses.

9. The researchers report that the two groups were similar in age and education. In your opinion, is this important? Explain.

10. In their research article, the researchers refer to their study as a "preliminary study." In your opinion, what aspects of the study, if any, qualify it as preliminary?

11. In their research article, the researchers indicated that they chose the paradigm of male interns gambling because it had been used in previous studies of other types of populations (e.g., heroin addicts). In your opinion, is this a good reason to select this paradigm?

12. Are the results of this study interesting or surprising to you? Explain.

Exercise 42   Effectiveness of an AIDS Prevention Program
Wilcoxon Matched-Pairs Test

Statistical Guide

The Wilcoxon Matched-Pairs test determines the significance of the difference between two sets of ranks obtained from matched pairs. For example, a person's rank on a pretest (before treatment) may be matched with the same person's rank on a posttest (after treatment).

Like other significance tests, when the Wilcoxon Matched-Pairs test indicates that p is equal to or less than .05, we usually declare the difference to be statistically significant (i.e., reject the null hypothesis). The smaller the value of p, the more significant the difference.

Excerpt from the Research Article[1]

SECRETS, an AIDS prevention program, was a theatrical production performed by young actors portraying adolescents…. SECRETS provided HIV/AIDS information and role modeled HIV risk-reduction behaviors.

Data on sexual risk-taking behavior were collected prior to and 3 months after students attended the AIDS prevention program. Sexual risk-taking behavior was considered one variable by summing the frequency of six behaviors [such as number of times a student engaged in sexual intercourse without using a condom and number of sexual partners]. The responses were from 0 to 5 for each behavior with a potential range from 0 to 30.

The sample, at pretest, had a relatively low level of sexual risk-taking behavior. The mean sexual risk-taking score was 1.8 ($SD = 3.5$)…. The sexual risk-taking scores ranged from 0 to 19 for this sample.

As the majority had not had sexual intercourse and there was a great variance in sexual behavior, the sample was categorized as low (≤ 1.8) and high (>1.8) sexual risk-taking groups…. The Wilcoxon Matched-Pairs test was performed for the low- and high-risk sexual risk-taking groups.

There was a significant difference between the pre- and posttest sexual risk-taking behavior for the low-risk group, Wilcoxon (129) = –4.9, $p = .0000$. For the low-risk group, 100 out of 127 (79%) had lower scores at posttest than at pretest.

There was also a significant difference between pre- and posttest sexual risk-taking behavior for the high-risk group, Wilcoxon (52) = –2.1, $p = .04$. For the high-risk group, 32 out of 52 (62%) had lower scores at posttest than at pretest.

The findings of the study suggested support for SECRETS in decreasing sexual risk-taking behavior among high school students.

---

[1] Source: Hanna, K. M., Hanrahan, S., Hershey, J., & Greer, D. (1997). Evaluation of the effect of an AIDS prevention program on high school students' sexual risk-taking behavior. *Issues in Comprehensive Pediatric Nursing, 20,* 25–34. Copyright © Taylor and Francis, 1997. Used with permission.

Questions for Exercise 42

Part A: Factual Questions

1. For all students, what was the average pretest score?

2. Is the distribution for all students on the pretest skewed? If yes, is it a positive or negative skew? (Hint: Consider the mean and the range.)

3. The researchers state that "there was great variance in sexual behavior." What specific statistics were reported that give more information on this matter?

4. A student with a score on the pretest of 2 would have been classified as belonging to which group?

5. For the low-risk group, should the null hypothesis be rejected? Explain.

6. For the high-risk group, should the null hypothesis be rejected? Explain.

7. For which of the tests was the level of significance higher? Explain the basis for your choice.
   A. The one for the high-risk group
   B. The one for the low-risk group

8. Was the test for the low-risk group statistically significant at the .001 level? Explain.

9. Was the difference for the high-risk group statistically significant at the .05 level? Explain.

Part B: Questions for Discussion

10. The researchers reported the average pretest score for all students but not the corresponding posttest score. Speculate on why they did not report the latter.

11. Was there a control group in this study? Explain.

12. In their article, the researchers mention that the sample included only students who had consented to be in the study and also had parental consent for them to participate. Consent was provided by 36% of parents and participants. Is this a limitation of the study? Explain.

13. Speculate on why the researchers used the Wilcoxon Matched-Pairs test instead of a *t* test or ANOVA.